EARTH FOOD SPIRULINA

How this remarkable blue-green algae can transform your health and our planet.

D1357771

New updated and revised edition

Robert Henrikson
Ronore Enterprises, Inc., Kenwood, California

Earth Food *Spirulina*
How this remarkable blue-green algae can transform your health and our planet.

Copyright © 1989, 1994, 1997, by Robert Henrikson.
All Rights Reserved.
No part of this book may be copied or reproduced in any form without the written permission of the publisher.

Cover design collage of Earthrise Farms by Chris Yoro and Rodney Ruppert.

Back cover photo of Robert Henrikson courtesy of Uni-Bond International, Guangzhou, China.

Library of Congress Catalog Card Number: 89-091683
ISBN 0-9623111-0-3

Printed in the United States of America
Published by
Ronore Enterprises, Inc.
PO Box 1188
Kenwood, California 95452 USA

First Printing, June 1989.
Second Printing, October 1989.
Third Printing, Revised Edition, April 1994.
Fourth Printing, Revised Edition, January 1997.

Acknowledgements

I wish to express my appreciation for the dedicated people who have pursued the dream of personal and planetary transformation by developing spirulina and other microalgae.

I would like to thank Larry Switzer, David Donnelley, Shigekuni Kawamura, Alan Jassby, Robert Bellows, Takemitsu Takahashi, Hidenori Shimamatsu, Terry Cohen, Bill Lackey, Yoshimichi Ota, Amha Belay, Juan Chavez, Ron Henson, Ernest Thomas, Paul Schofield, Mark Belock and many other people who have contributed.

For their special assistance in editing this manuscript, I wish to thank my brothers John Henrikson, Richard Henrikson and Jack Wight, and for assistance in revising this edition, Rodney Ruppert and Chris Yoro.

Sunset over the world's largest spirulina farm,
Earthrise Farms in the sunny California desert.

Table of Contents

Foreword

This book completes another step in the realization of a shared vision quest. Robert and I have been "Brothers of a Great Dream" for more than fifteen years. It is a dream no one owns but all can share – if they have the imagination and faith to rise to its promise, to know that it can and must come true. It is a Great Dream because we intuit that it is also the dream of the spirit that creates our destiny – the spirit of this living world. Our instincts tell us that this living world is not only the home of life, but is itself a gigantic, self-evolving organism some have called Gaia. She too has embarked upon a four billion year vision quest: to fill herself with conscious life. We humans are an essential player in her Great Dream. In our awakening to her reality, we sense that she too is further awakening. Somehow Gaia herself is coming to fuller consciousness of her embryos of awareness called humans. Such is the true position of humanity – awakening to the full miracle of its most ancient and living parent, this living, conscious Earth.

If we fully accept the fact of her miracle, the fulfillment of her vision quest becomes ours. Our awakening invites us to care for our most ancient and sacred parent by co-creating upon her and with her. We are coming to realize that we can no more ravage and destroy her and survive than a foolish child can ravage and destroy its own mother. She asks us to honor her by honoring each other and the preciousness of life itself. And if we respond to her, we can be certain that she in turn will continue to be generous with us and with our far descendants. We can awake to the startling good news that we can live and thrive within her miraculous being, that we can enjoy a sustainable sanity, peace, and abundance which until now has only been glimpsed by visionaries. This is how her vision quest is fulfilled in our awakening.

I honor this book because I know it was created in the service of her spirit. It reveals and explains one of the vital gifts the parent – Earth – offers its awakening child – humanity.
– *Larry Switzer, May 1989.*

Robert Henrikson and Larry Switzer.

Invocation

Spirulina speaks to the human species, on behalf of the first species – algae.

"For a billion years we filled the Earth's atmosphere with enough oxygen so new lifeforms could evolve. We observed and participated in the unfolding of the diverse lifeforms on this planet for billions of years more. Paradise unfolded and we loved it.

Just moments ago, your species appeared. In the past 50 years, you humans have been shutting down the life support systems of our planet. Your unusual ability to upset the biosphere, deplete the ozone layer, increase global warming, deforest the land, expand deserts and pollute land, water and air has aroused our attention.

We do not depend on your gratitude for having provided you a beautiful planet, but a little cooperation is in order. Your planetary plundering destroys the opportunity for life to unfold fully in all its forms. If you persist, your species will likely perish too. We algae will survive, and over the eons will again foster new life.

We do recognize, however, your unique destiny on this planet. You are the most immature, yet the most intriguing of all species. The future of planetary evolution rests with you. Your survival and evolution require healing our planet in the next 20 years.

Heal yourself within, heal your relationships with your own species, and heal our planet. From this great challenge will emerge your highest creativity. Taking this evolutionary jump, you will begin some very interesting things here, and we want to participate in them with you.

We offer our wisdom for your personal and planetary health. Embrace, befriend and learn from us. Rediscover the ancient wisdom of you biological ancestors."

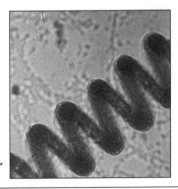

Spirulina, "little spiral."

3

Introduction to a perfect food to restore our health

Our heady plunge into technology has upset the balance of Earth's biosphere. Our consumer lifestyle hastens the deterioration of our environment. We transform Earth's resources into trash and pollution at a faster and faster pace. We are literally consuming Earth.

Emerging from ecosystem breakdown are super bacteria, viruses and chemical toxins that threaten our own health. At the same time, a portion of the Earth's population is seeking super health to boost their immune system, resist disease and retard the aging process.

All this arrives synchronistically at the end of the millennium, when we sense a new age approaching: an age of information, an age of transformation, or an age of light.

How can we prepare ourselves and make those choices that restore our personal health, the health of our society and the entire planet?

Food choices for health restoration

1. Does our food restore our personal health?

Many foods we eat are toxic to long term health. They lead to cancer, heart and degenerative disease, immune breakdown and chronic lethargy. Two-thirds of all deaths in the USA are attributed to diet.

2. Does our food restore our human species?

Much of the global food production and distribution system creates hunger in a world of abundance. Two-thirds of humanity live in poverty and scarcity. Over 40,000 children die of malnutrition and related diseases every day.

3. Does our food restore our planetary ecosystem and its remarkable biodiversity?

About three-quarters of all fertile land in the temperate and tropic zones is devoted to agriculture. Our world economy depletes fertile soil, wastes fresh water, pollutes the environment, kills other living species, turns rainforest into desert, cropland into wasteland. The unsustainable way we produce our food may represent the greatest threat to the health of our biosphere.

Both planetary and personal health are under environmental stress

This inefficient expropriation of resources increases scarcity, poverty and suffering for two-thirds of the world's people. Agribusiness desertifies the planet, eliminates biodiversity, pushes global warming and steals resources from future generations.

Environmental stress accelerates the breakdown of Earth's ecosystems. These systems cannot adjust as quickly to the changes humans have recently introduced. Great bodies of water are dying from pollution. Forests around the world are dying as if they were suffering from some immune system collapse.

Neither can human bodies quickly adjust to these rapid changes. Increased ultraviolet radiation from the depletion of the ozone layer causes skin cancer, cataracts and suppression of our immune system. Simultaneously, immune system diseases like AIDS have appeared.

Introduction

Antibiotics are no longer effective against new resistant bacteria. *"The End of Antibiotics"* (Newsweek March 28, 1994) claimed "the rise of drug resistant germs is unparalleled in recorded biologic history." Over prescription of antibiotics in the last 40 years has accelerated the mutation of resistant bacteria. People taking medications and antibiotics are more vulnerable. Antibiotics kill beneficial flora in the intestines, making room for infection by drug resistant bacteria.

Much more infectious mutant retro viruses than AIDS are coming. *"The Danger Zone"* (May 4, 1994) broadcast by *CBS 48 Hours* and new books like *"The Coming Plague* – newly emerging diseases in a world out of balance" warn about these deadly viruses. The sensational movie *Outbreak* (1995) used a mutant airborne ebola virus to scare us. *Hantavirus* carried by rats and new viruses emerging from the tropical rainforests are examples of infectious airborne viruses.

Although this situation seems pretty grim, there is good news. A remarkable paradigm shift is already underway.

The search for designer foods to restore health

In animal nutrition, as antibiotics are becoming ineffective, scientists are replacing them with *probiotics,* special therapeutic foods that boost the immune system and resistance to disease. In human medical research, scientists are rushing to identify probiotic foods that enhance our immune system. Researchers are scouring the globe for diverse new foods and plants for cancer preventing compounds. The goal of the National Cancer Institute Experimental Food Program is to study, assess and develop experimental, or *designer foods,* rich in disease and cancer preventing substances.

There's a revolution in health: We've been hearing about *antioxidants, nutraceuticals,* and *designer foods* loaded with *functional nutrients*: the leading edge of nutritional research in the 1990s. Another *Newsweek* story (April 25, 1994) proclaimed: *"Better than Vitamins: can phytochemicals prevent cancer?"* The article extolled remarkable compounds in whole foods that can prevent disease, even better than isolated vitamins and nutrients.

Probiotics, nutraceuticals, phytochemicals, designer foods – where do we start? With the *original food designed by nature – spirulina!* To raise your energetic frequency, bring this light food into your life.

The perfect food
to restore our health and our planet

The first photosynthetic lifeform was designed by nature 3.6 billion years ago. Blue-green algae, cyanobacteria, is the evolutionary bridge between bacteria and green plants. It contained within it everything life needed to evolve. This immortal plant has renewed itself for billions of years, and has presented itself to us in the last 15 years. Spirulina has 3.6 billion years of evolutionary wisdom coded in its DNA.

Does spirulina contain antioxidants? Yes. Is it a probiotic food? You bet! Is it a nutraceutical? That too. Is it loaded with phytochemicals? All kinds. It contains compounds like phycocyanin, polysaccharides, and sulfolipids that enhance the immune system. This superfood has the most remarkable concentration of functional nutrients ever known in any food, plant, grain or herb.

On top of this, spirulina delivers more nutrition per acre than any other food on the planet. This has extraordinary implications for more efficient and less damaging food production for the future.

Each day new research brings to light the wonders (hidden) in microscopic algae. In 1989, the National Cancer Institute announced sulfolipids extracted from blue-green algae were 'remarkably active' against the AIDS virus in test tube experiments. Sulfolipids can prevent viruses from either attaching to or penetrating into cells, thus preventing viral infection.

Research published from 1991-96 has shown phycocyanin and polysaccharide extracts of spirulina increase macrophage production, bone marrow reproduction, strengthen the immune system and disease resistance in fish, mice, chickens, cats and human cells.

Algae is in its infancy as a food, medicine and biochemical resource. Spirulina, a descendant of Earth's first photosynthetic life form, was rediscovered about 30 years ago. Just 15 years ago, it burst into public awareness as a powerful new food with a promise as a food source to help feed the world's people.

This magnificent idea caught our imagination. Compared to the five billion years of Earth history, or to the millions of years of humanity, or to the thousands of years of human food development, 15 years is only an instant in time.

Cultivation of grains and development of irrigation took thousands of years. Soybeans, a newcomer, took 50 years to emerge from obscurity. The last 15 years progress in algae technology is astounding.

The first chapter (1) looks at the role of algae in history and the implications from its productivity. How can spirulina transform your personal health? (2) Its nutritional attributes are described and compared with other foods. (3) A review of personal self care programs shows how to use and benefit from this superfood. (4) The extensive clinical research suggests spirulina is a probiotic and therapeutic food.

What is the role of spirulina in global economics and politics? (5) The variety of products around the world is probably more than you realize. (6) Ecological technology is used to cultivate spirulina, and future growing systems will produce exciting new products. (7) A critical look shows the resource advantages of spirulina production.

How can spirulina help restore our planet? (8) Farms large and small are blossoming in the developing world. (9) Looking to the future, spirulina and other algae are featured in projects to restore and regreen the face of our planet.

Just as our body is composed of billions of cells working together as a single being, billions of lifeforms on Earth are working together as one living organism. By adopting ecological food choices for restoring our own health, we help restore humanity and our planet.

Algae harvests sunlight. It transforms light to living matter more efficiently than other plants. Eat the light in spirulina and bring light into your own cells. Eat lighter, eat less. Raise your energy level to embrace the pace of change in this age of transformation. Consume less, live lighter on the Earth. Participate in the unfolding story of earth consciousness rising.

The oldest organisms – the ones who gave us life – are back.

They represent some of the many solutions for restoring our planet in the next 20 years. Perhaps for this reason algae have arrived in our consciousness. Spirulina is one of the best known. This is the story of its past and present, and a glimpse into its future.

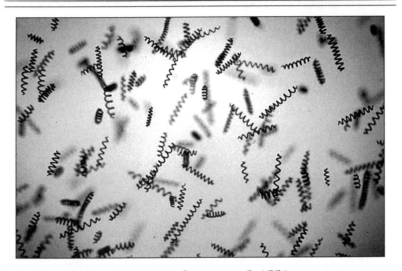

Chapter 1

Rediscovery of a 3.5 billion year old immortal lifeform

Spirulina is the immortal descendent of the first photo-synthetic lifeform. Beginning 3.5 billion years ago, blue-green algae created our oxygen atmosphere so other life could evolve. Since then, algae have helped regulate our planet's biosphere.

Algae are two-thirds of the Earth's biomass. Thousands of algal species covering the Earth are now being identified for food, pharmaceuticals, biochemicals and fertilizers. Algae represent one of the solutions we need to produce food while restoring our planet.

- In the Beginning ... were blue-green algae.
- Thousands of algal species cover the Earth.
- Algae through human history.
- Rediscovery of human use – Kanembu and Aztecs.
- Spirulina lakes and pink flamingos.
- A new era of ecological agriculture.

In the beginning were blue-green algae

When life began on Earth, the carbon dioxide level in our atmosphere was probably 100 times greater than it is today. Life began in a greenhouse atmosphere, and microalgae played the central role in transforming this inhospitable planet into the beauty and richness that makes up life today. How this occurred is particularly relevant in view of our concern with global warming.

Scientists believe the Earth formed 4.5 billion years ago, and the first lifeforms appeared 3.6 billion years ago. There is considerable controversy about how life was actually created on this planet.

One theory, growing in support over the past decade, asserts the Earth is a self-regulating, living organism, actively maintained by lifeforms on its surface. In *The Ages of Gaia*, James Lovelock offers an intriguing description of how lifeforms on the planet's surface modified and regulated the atmospheric gas composition as life evolved. Because the young sun was 25% cooler at the beginning of life, the greenhouse effect kept a cooler planet warmer. Earth's nitrogen atmosphere, without any oxygen, was rich in greenhouse gases (like carbon dioxide and methane) absorbing and trapping radiant heat, with the infrared radiation rising from the surface. The oceans were filled with iron, sulfur and other compounds in solution because there was no free oxygen. These substances reacted with and removed oxygen, so the Earth had a great capacity to prevent the appearance of free oxygen.[1]

The first living bacteria, the procaryotes, consumed chemical nutrients as food, but some adapted the energy of the sun to make their own food. The first photosynthesizing procaryotes, called cyanobacteria or blue-green algae, used light energy to break apart the abundant carbon dioxide and water molecules into carbon food compounds, releasing free oxygen. Fossils dating back 3.6 billion years, show filaments of these single cells stacked end on end. The shape unmistakably resembles spirulina.

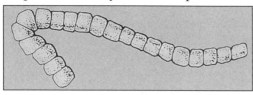

1.2.
Drawing of a 3.6 billion year old cyanobacteria fossil.

 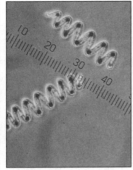 

1.3. Long filaments. **1.4. Perfect spiral coils.** **1.5. Electron microscope.**

Views of spirulina under the microscope.

Iron and sulfur compounds in the oceans mopped up almost all of the free oxygen immediately. Methanogen bacteria consumed decomposed algae and converted the carbon in it to methane gas and carbon dioxide, compensating for the removal of carbon dioxide by photosynthesizing algae.

Lovelock describes the planet during this period as a brownish red hazy planet, with a layer of methane smog in the atmosphere, offering similar protection as the ozone layer today. The cyanobacteria colonized the oceans and formed a thin film on the land masses.[2] These blue-green algae carried their genetic information in DNA strands in the cell membrane and could exchange information by exchanging plasmids with another. In this way, the organism became essentially immortal.

"The Earth's operating system was populated totally by bacteria. It was a long period when the living constituents of Gaia could be truly considered a single tissue. Bacteria can readily exchange information, as messages encoded on low molecular weight chains of nucleic acids called plasmids. All life on Earth was then linked by a slow but precise communication network."[3]

Over a billion years passed. When the oxygen absorbing compounds in the oceans were used up, the atmospheric concentration of oxygen increased rapidly. Methanogen bacteria retreated into the only environments devoid of oxygen – beneath the sea floor, in marshes, and in the guts of other organisms.

About 2.3 billion years ago, a new period began when oxygen may have reached a 1% level, and methane, a greenhouse gas, disappeared from the atmosphere, cooling the planet.[4]

Cells with nuclei appeared. This more powerful and complicated lifeform was supported by the higher oxygen concentration. These eukaryotes, such as microscopic green algae, may have formed from communities of individual bacteria living within an outer membrane of one of them. The nucleus contained organelles such as chloroplasts, the green bodies which photosynthesize. Because each organelle carried different genetic codes, the loss of information of one of them could mean the death of the cell. To overcome this possibility of death, sex evolved as a way to transfer information between cells.[5]

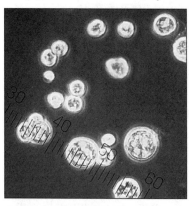

1.6. Chlorella, green microalgae with nucleus and strong cell walls.

About 600 million years ago, Earth entered the present phase with the evolution of large plants and animals. The power requirements of larger organisms like trees and dinosaurs needed a higher oxygen concentration, which increased and remained steady at 21%. For hundreds of millions of years, the Earth's biosystem has kept the oxygen level carefully balanced between 15%, where higher life forms cannot survive, and 25%, where forests would spontaneously combust in a global fire.

The procaryotes, cyanobacteria, or blue-green algae, still cover the land and water surfaces, part of the living mechanism for regulating the planet's biosphere. Our rediscovery and interest in this original lifeform is no accident. It represents our need to return to the origins of life to understand and heal our planet. Realizing that algae took billions of years to build and maintain the atmosphere, it is remarkable that humanity has raised the carbon dioxide concentration over 25% in merely one hundred years.

How important is the contribution this original lifeform? Brian Swimme, in *The Universe is a Green Dragon* writes:

"I think we should take the procaryote as the mascot of the emerging era of the Earth. What better organism to symbolize the vast mystery of the Earth's embryogenesis ... Let's just hope we can emulate some of the achievements of the procaryotes ... To begin with, it would be wonderful if we could contribute something as essential to Earth's life as oxygen."[6]

Thousands of algal species cover the earth

There may be more than 25,000 species of algae, living everywhere. They range in size from a single cell to giant kelp over 150 feet long. Most algae live off sunlight through photosynthesis, but some live off organic matter like bacteria.

Larger algae, like seaweeds, are macroalgae. They already have an important economic role. About 70 species are used for food, food additives, animal feed, fertilizers and biochemicals.

Microalgae can only be seen under a microscope. Some serve a vital role for breaking down sewage, improving soil structure and fertility and generating methane and fuels for energy. Others are grown for animal and aquaculture feeds, human foods, biochemicals and pharmaceuticals.

Microalgae in the ocean, called phytoplankton, are the base of the food chain and support all higher life. The rich upwelling of nutrients caused by the major currents meeting the continental shelf, or nutrients from river basins sustain phytoplankton growth.

There are blue-green microalgae like spirulina and aphanizomenon, green algae like chlorella and scenedesmus, red algae like dunaliella, and also brown, purple, pink, yellow and black microalgae. They are everywhere – in water, in soils, on rocks, on plants. Blue-green algae are the most primitive, and contain no nucleus or chloroplast. Their cell walls evolved before cellulose, and are composed of soft mucopolysaccharides. Blue-green algae do not sexually reproduce; they simply divide.

Some blue-green algae can fix atmospheric nitrogen into organic forms. This is very important because organic nitrogen is essential for building proteins and amino acid complexes in plants and animals. Although nitrogen gas comprises 78% of the atmosphere, it is not usable by most plants and animals. For more productive crops, nitrogen must be added to soils. Organic nitrogen can only come from adding chemical fertilizers, from existing microbial mineralization of organic matter, by nitrogen-fixing bacteria in legume roots, or by nitrogen-fixing blue-green algae.

Because of this ability to fix nitrogen, blue-green algae is often the first lifeform to colonize a desolate land area – in deserts, in volcanic rocks, on coral reefs, and even in polar regions, working with lichen to fix nitrogen to the rocks to begin life in the tundra.[7]

Nitrogen-fixing blue-green algae are being developed as natural biofertilizers, but they are not always safe to eat. Many kinds of microcystis, anabaena and aphanizomenon are toxic just like some mushrooms and land plants. Harvesting wild blue green algae from lakes presents a risk of contamination by algal toxins.

Spirulina, whose scientific name is *arthrospira*, is an edible, non-nitrogen fixing blue-green algae. with a long history of safe human consumption and over 30 years of safety testing. It meets all international food quality and safety standards. Specially designed farms where spirulina is cultivated under controlled conditions, do not allow the growth of other contaminant blue-green algae, as in lakes and waterways.

Algae in human history

Microalgae have kept a rather low profile, but their interaction with humans is notable on several occasions. The Bible describes when the Israelites were starving in the wilderness, God provided 'manna' – a flake-like thing, lying on the ground. They gathered the manna and baked it into bread. Some believe the manna was a kind of lichen – a combination of fungus and blue-green algae that formed a crust on the rocks and ground.[8]

Another story took place a thousand years ago in Vietnam. A monk named Khong Minh Khong discovered rice was far more productive when a water fern, azolla, was planted in the paddies. The grateful farmers built temples to him after he died, but kept it secret.

Some 700 years later, a woman named Ba Heng rediscovered azolla. Growing rice with azolla continued for centuries, increasing yields and saving many people from starvation. Only this century did scientists discover blue-green algae living on the fern were fixing nitrogen as a natural biofertilizer for the rice.[9]

Although freshwater or inland algae has not been eaten nearly as much as larger marine seaweeds, a survey of historical literature revealed at least 25 separate cases where at least nine types of wild freshwater algae were collected and eaten in 15 countries.[10] This non-seaweed algae has been used in a variety of soups, spreads and sauces and may have been an important source of vitamins and minerals.

When microscopic algae could be easily collected because it formed into larger colonies of mats or globules, it played a culinary and therapeutic role similar to many higher plants. So, eating algae may have been limited only by the difficulty of collecting these tiny organisms.[11] Two cases, on separate continents, involved spirulina.

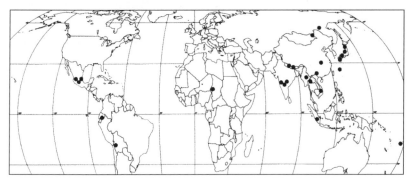

1.7. Locations of traditional human consumption of freshwater microalgae, excluding ocean seaweeds (courtesy Alan Jassby).

Rediscovery of the human use of spirulina

In 1940, a little known journal published a report by French phycologist Dangeard on a material called *dihé,* eaten by the Kanembu people near Lake Chad. *Dihé* is hardened cakes of sun-dried blue-green algae collected from the shores of small ponds around Lake Chad. Dangeard also heard this same algae populated a number of lakes in the Rift Valley of East Africa, and was the main food for the flamingos living around those lakes. His report went unnoticed.

Twenty five years later, a Belgian Trans-Saharan expedition discovered a blue-green material covering the waters around the shores of Lake Chad. A botanist with the expedition, Leonard, came across curious blue-green cakes in native markets of Fort Lamy (now Ndjemena) in Chad. When locals said these cakes came from areas near Lake Chad, Leonard recognized the connection between the algal blooms and dried cakes sold in the market.

1.8. Collecting spirulina from a lake in Chad (photo by J. Maley).
1.9. Spirulina cakes (dihé) on sale in a local market in Chad
(photo by J. Maley).

Desert winds pushed the mats of algae to the shores. Kanembu people collected the wet algae in clay pots, drained out the water through bags of cloth and spread out the algae in the sand to dry in the sun. When dry, women cut the algae cakes into small squares for sale in the local market. *Dihé* is crumbled and mixed with a sauce of tomatoes and peppers, and poured over millet, beans, fish or meat. It is eaten by the Kanembu in 70% of their meals. Pregnant women eat *dihé* cakes directly because they believe its dark color will screen their unborn baby from the eyes of sorcerers.[12] Spirulina is also applied externally as a poultice for treating certain diseases.

1.10. Kanembu women gathering spirulina from area around Lake Chad. Drawing in <u>Human Nature</u>, March 1978 (article by Peter T. Furst).

At the same time, a company director in Mexico read about spirulina and realized it was the same algae clogging the soda extraction plant on Lake Texcoco. Although spirulina was not then eaten as a food in Mexico, an historical literature search revealed it was harvested, dried, and sold for human consumption 400 years earlier, at the time of the Spanish conquest.

Spanish chroniclers described fisherman with fine nets collecting this blue colored *'techuitlatl'* from the lagoons, as seen in Figure 1-10, and making bread or cheese from it. Other legends say Aztec messenger runners took spirulina on their marathons. *Techuitlatl* was mentioned by naturalists until the end of the 16th century, but not after that. Probably, it disappeared soon after the Spanish conquest. The great lakes in the Valley of Mexico were drained to make way for the new civilization. The only remnant today, Lake Texcoco, still has an living algae culture.

1.11.
Aztecs harvesting blue-green algae from lakes in the Valley of Mexico. Drawing in Human Nature, March 1978.
(article by Peter T. Furst).

Spirulina lakes and pink flamingos

Besides Lake Texcoco, the largest spirulina lakes are in Central Africa around Lake Chad, and in East Africa along the Great Rift Valley. Under normal water conditions, spirulina may be one of many algal species. But the more alkaline and salty the water becomes, the more inhospitable it becomes to other lifeforms, allowing it to flourish as a single species.

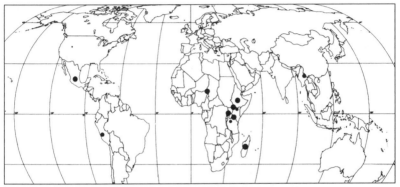

1.12. Lakes with natural spirulina blooms.

Lakes Bodou and Rombou in Chad have a stable monoculture of spirulina dating back centuries. It is a major species in Kenya's lakes Nakuru and Elementeita and Ethiopia's lakes Aranguadi and Kilotes. The lesser flamingo evolved a filter in its beak especially to eat spirulina. Millions of flamingos feed entirely on algae when it is abundant.

1.13. Pink flamingos in an African lake feeding on spirulina.

Spirulina thrives in alkaline lakes where it is difficult or impossible for other microorganisms to survive. Because the bacteria level in alkaline water is quite low, the bacteria count in spirulina, harvested and dried without further processing, is insignificant.[13] Algae pioneers have dreamed of harvesting from these lakes to feed the millions of nearby hungry people now suffering from chronic hunger.

In natural lakes, the limited supply of nutrients usually regulates growth cycles. New nutrients come from either an upwelling from inside the earth, when rains wash soils into the lakes, or from pollution. The algae population grows rapidly, reaches a maximum density, and then dies off when nutrients are exhausted. A new seasonal cycle begins when decomposed algae release their nutrients or when more nutrients flow into the lake

Algae cultivation is an evolutionary step in agriculture

Over thousands of years, humans have dramatically increased food productivity, at progressively greater environmental costs. Domesticating plants and animals encouraged the first permanent human settlements. About 7000 years ago, irrigation brought water to the land and subsequent food surpluses supported the first great river valley civilizations. Thousands of years later, when the land salted up from over-irrigation, these civilizations vanished.

A thousand years ago, the invention of an efficient plough in Europe allowed easier tilling of the soil. Europeans cut down the vast original forests bringing new areas under cultivation and new prosperity to the continent. The 19th century industrial revolution introduced mechanized agriculture, climaxing in the so-called 'Green Revolution' exported from the United States in the 1960s and 1970s.

Modern agriculture has boosted short term productivity by using seed hybrids and massive fertilizer, pesticide, water and energy inputs. Productivity has been achieved by simply ignoring many hidden costs, such as the consumption of non-renewable fossil fuels for fertilizers and machinery, pollution of soil and water through excessive use of chemical fertilizers, and depletion of soils. This ecological damage will be paid by future generations.

Successful algae cultivation requires a more ecological approach to begin with. A pond of spirulina is a living culture and the whole system, not just a few inputs, must be considered. If one factor changes, the entire pond environment changes – quickly. Because algae grows so fast, the result can been seen in hours or days, not seasons or years like in conventional agriculture.

Algae scientists talk of 'balancing pond ecology' for sustainable growth. Pesticides and herbicides would kill many microscopic life forms in a pond, so algae scientists have learned how to balance the pond ecology to keep out weed algae and zooplankton algae eaters without using pesticides or herbicides.

Ecological food production is the next stage in agriculture. This represents both an increase in productivity and stewardship of the Earth's resources. Organic and biodynamic farming methods, permaculture, aquaculture and low tillage farming are practices now becoming more popular. Algae cultivation is a new addition to ecological food production.

The hope of spirulina

A spirulina farm is an environmentally sound green food machine. Cultivated in shallow ponds, this algae can double its biomass every 2 to 5 days. This productivity breakthrough yields over 20 times more protein than soybeans on the same area, 40 times corn and 400 times beef. Spirulina can flourish in ponds of brackish or alkaline water built on already unfertile land. In this way, it can augment the food supply not by clearing the disappearing rainforests, but by cultivating the expanding deserts.

In one of the first books on microalgae, *Spirulina, the Whole Food Revolution,* Larry Switzer wrote:

"For the first time since the appearance of man, both wilderness and food productivity can be increased simultaneously with a new technology. This is a choice that man has never had before. The rediscovery of this ancient life as a human food has great implications for us all, now and in the 21st century. It is an example of the myriad of unexpected and astounding solutions to basic world problems that are now beginning to appear together on this planet."

Chapter 2

Which came first the Chicken or the Egg?

The answer is **SPIRULINA**

Over 3.5 billion years old, Spirulina is the Earth's oldest source of life enhancing phytonutrients, antioxidants and protein rich nutritiion.

One of the World's Healthiest Foods.
Ecologically grown at Earthrise Farms in California.

A nutrient rich super food for super health

Our modern diet is filled with depleted, over-processed convenience foods. Many people supplement with extra vitamins and minerals. Now science is looking beyond vitamins to *"Phytonutrients."*

"It is whole foods that pack the disease preventing wallop. That's because they harbor a whole ratatouille of compounds that have never seen the inside of a vitamin bottle." (*Newsweek*, April 25, 1994).

Spirulina called a superfood because its nutrient profile is more potent than any other food, plant, grain or herb. These nutrients and phytonutrients make spirulina a whole food alternative to isolated vitamin supplements.

- Protein and Amino Acids.
- Vitamins and Minerals.
- Essential Fatty Acids.
- Phytonutrients.
- Comparing Green Superfoods.

Concentrated green super food

Early research documented spirulina's safe consumption by traditional peoples. When scientists discovered that spirulina grew so fast, yielding 20 times more protein per acre than soybeans, they named it a *food of the future*. Spirulina is the best vegetable protein source, with a protein content of 65%, higher than any other natural food. Yet, an even greater value is found in its concentration of vitamins, minerals and other unusual nutrients.

Three to ten grams a day delivers impressive amounts of beta carotene, vitamin B-12 and B complex, iron, essential trace minerals, and gamma-linolenic acid. Beyond vitamins and minerals, spirulina is rich in phytonutrients and functional nutrients that demonstrate a positive effect on health. For undernourished people in the developing world, spirulina brings quick recovery from malnutrition. In Western overfed food culture loaded with unhealthy and depleted foods, spirulina can renourish our bodies and renew our health.

It is legally approved as a food or food supplement in Europe, Japan and many other countries around the globe. The United States Food and Drug Administration confirmed in 1981 that spirulina is a source of protein and contains various vitamins and minerals and may be legally marketed as a food supplement.[1] Many countries have set up food quality and safety standards for spirulina.

Nutritional depletion of modern foods

Today's food is lower in essential nutrients than foods produced only 50 years ago. Farming practices have depleted our soils of minerals. Microorganisms in the soil contributing the valuable mineral content are declining because the overuse of chemical fertilizers destroys these microorganisms. Agribusiness chooses hybrid strains based on harvestability, appearance, and storageability, rather than nutrient content. Furthermore, the long shipping and storage time between harvest and selling reduces nutrient content.

At the same time, researchers say increased stress from environmental pollutants and lifestyle demands have increased our dietary requirements for certain essential nutrients.[2] As a result, many of us do not trust the quality of our foods. Twenty years ago only health food consumers used nutritional supplements. Today, at least some supplements are used by almost everyone.

Beyond isolated vitamins and minerals

Vitamins and minerals in foods are bound to natural food complexes with proteins, carbohydrates and lipids. The human body recognizes this entire food complex as food. Most supplements, however, are synthetic combinations of isolated USP vitamins and minerals. These are often formulated to claim 100% of the Daily Value (DV) on labels. But these vitamins and minerals are not bound to anything, and may have an entirely different chemical structure than those found in foods.

Formulas may ignore antagonistic and synergistic effects of vitamins and minerals both in regard to absorption and metabolic reactions once absorbed. Complex factors in whole foods that aid absorption, such as chelating agents, may be missing in laboratory formulated vitamins and minerals. It is well known that many supplements, especially calcium and iron, are not well absorbed.

Supplement megadoses were attempts to overcome absorption problems with an approach that more is better. Unfortunately, this may not be true. Absorption of vitamins and minerals is limited by uptake mechanisms in the intestines, and megadoses are largely excreted. Some say megadosing is more than a waste of money, it is unwise. If the body relies on formulated supplements, it might get lazy or 'forget' to extract nutrients from foods efficiently.

Next, clever technicians invented chemical chelators, transporters and time release agents to make mineral supplements better absorbed. In other attempts, vitamins and minerals have been extracted from some food sources. These concentrates may have toxic residues if chemical solvents are used to remove them. Recently touted are 'food-form' type supplements. In this case, USP vitamins and minerals are recombined in a vat with yeast bacteria. It is claimed these products are 'just like food'.

Most people believe it is better to get nutrients from natural foods. Since many conventional foods are nutrient depleted, more people are taking spirulina and other green superfoods. These whole foods offer functional nutrients and phytochemicals, new frontiers for disease prevention research, way beyond isolated vitamin and mineral supplements.

Protein and amino acids

The building blocks of life are protein and amino acids. When comparing protein sources, several criteria should be considered: protein quantity, amino acid quality, usable protein, digestibility, and the negative 'side effects' from fat, calorie and cholesterol content.

Protein: Spirulina has the highest protein of any natural food (65%); far more than animal and fish flesh (15-25%), soybeans (35%), dried milk (35%), peanuts (25%), eggs (12%), grains (8-14%) or whole milk (3%).

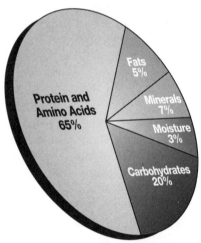

2.2. Composition of spirulina.

Amino Acid Quality: Protein is composed of amino acids. Essential amino acids cannot be manufactured in the body and must be supplied in the diet. Non-essential amino acids are needed too, but the body can synthesize them. Essential amino acids, plus sufficient nitrogen in foods, are needed to synthesize the non-essential amino acids. A protein is considered complete if it has all the essential amino acids. Spirulina is just that, a complete protein.

The body requires amino acids in specific proportions. If a food is low in one or more amino acids, those amino acids are called limiting amino acids, and the body cannot use all the amino acids completely. The most ideal proportion of amino acids is found in eggs. All other foods have some limiting amino acids.

2.3. Spirulina amino acid composition[a]

Essential Amino Acids	per 10 grams		% total
Isoleucine	350	mg	5.6 %
Leucine	540	mg	8.7 %
Lysine	290	mg	4.7 %
Methionine	140	mg	2.3 %
Phenylalanine	280	mg	4.5 %
Threonine	320	mg	5.2 %
Tryptophan	90	mg	1.5 %
Valine	400	mg	6.5 %
Non-Essential Amino Acids	**per 10 grams**		**% total**
Alanine	470	mg	7.6 %
Arginine	430	mg	6.9 %
Aspartic Acid	610	mg	9.8 %
Cystine	60	mg	1.0 %
Glutamic Acid	910	mg	14.6 %
Glycine	320	mg	5.2 %
Histidine	100	mg	1.6 %
Proline	270	mg	4.3 %
Serine	320	mg	5.2 %
Tyrosine	300	mg	4.8 %
Total Amino Acids	**6200**	**mg**	**100.0 %**

a. Earthrise Farms, 1995.

Limiting amino acids in spirulina are methionine and cystine, but it is still higher in these amino acids than grains, seeds, vegetables and legumes, and higher in lysine than all vegetables except legumes. Spirulina complements vegetable protein and increases the amino acid quality if eaten within several hours of other foods. Over 100% of the daily essential amino acid requirements for a typical adult male are supplied by using only 36 grams of spirulina, about 4 heaping tablespoons (2.4).

Net protein utilization and usable protein

Feeding tests rank proteins by Net Protein Utilization (NPU) value, determined by amino acid quality, digestibility (proportion absorbed by the intestines) and biological value (proportion retained by the body). Dried eggs (94) have the highest value, followed by milk (70-82), fish (80) and meat (67). Spirulina (62) is similar to grains and has a higher NPU than nuts (2.5).

2.4. Adult essential amino acid (EAA) requirements provided by spirulina[a]

EAA	requirement g / day	spirulina gm / 10g	spirulina %/ 10g
Isoleucine	0.84	0.35	42 %
Leucine	1.12	0.54	48 %
Lysine	0.84	0.29	35 %
Methionine[b]	0.70	0.20	29 %
Phenylalanine[c]	1.12	0.58	52 %
Threonine	0.56	0.32	43 %
Valine	0.98	0.40	41 %

a. Jassby 1983, FNB 1975, Earthrise Farms 1995.
b. includes cystine.
c. includes tyrosine.

2.5. Protein quantity and quality for spirulina and other protein food sources[a]

Food	Protein %	NPU %	Usable Protein %
Spirulina[b]	65	62	40
Dried eggs, whole	47	94	44
Brewers yeast	45	50	23
Soy flour, whole	37	61	23
Dried milk, skim	36	82	30
Cheese, parmesan	36	70	25
Wheat germ	27	67	18
Peanuts	26	38	10
Chicken[c]	24	67	16
Fish[c]	22	80	18
Beef[c]	22	67	15
Sesame seed	19	60	11
Oats, whole flour	15	66	10
Wheat, whole flour[c]	14	63	9
Tofu, moist	8	65	5
Brown rice	8	60	5

a. Switzer, The Whole Food Revolution, 1982, pg 21.
b. value for spirulina NPU from O. Ciferri, Spirulina.
c. values are for highest protein levels for these groups.

By multiplying protein quantity by the NPU, we determine the usable protein as a percentage of the food's composition. Spirulina is second only to dried eggs, and higher than any of the common foods in the form in which they are usually purchased.

Protein digestibility is important for many people

Spirulina has no cellulose in its cell walls, being composed of soft mucopolysaccharides. This makes it easily digested and assimilated. It is 85 to 95% digestible. This easy digestibility is especially important for people suffering from intestinal malabsorption. Typically, many older people have difficulty digesting complex proteins, and are on restricted diets. They find spirulina's protein very easy to digest.

Spirulina is effective for victims of malnutrition diseases like kwashiorkor, where the ability of intestinal absorption has been damaged. Given to malnourished children, it is more effective than milk powders because milk's lactic acid can be difficult to absorb.

'Side effects' – fat, calories and cholesterol

Spirulina's fat content is only 5%, far lower than almost all other protein sources. Ten grams has only 36 calories and virtually no cholesterol. This means spirulina is a low-fat, low-calorie, cholesterol-free source of protein, and is not loaded with the fat, grease, calories and cholesterol of meat and dairy protein.

"One tablespoon (10 grams) of spirulina contains only 1.3 mg of cholesterol and 36 calories. In contrast, a large egg yields about 300 mg of cholesterol and 80 calories, while providing only the same amount of protein as the tablespoon of spirulina."[3]

People in the developed countries usually consume more than enough protein along with excessive fat, calories and cholesterol. Therefore, less dairy and meat protein sources are recommended.

Certain people need a higher protein intake, but not a corresponding increase in calories. For example, a typical pregnant woman is advised to increase her protein intake from 44 to 74 grams (68% increase) while increasing calories from 2000 to 2300 (15% increase). So additional protein must be low in calories to avoid an unnecessary weight increase. Spirulina is suitable in these cases where the 'calorie cost' is far lower than from dairy, meat and fish.[4]

Vitamins - protectors of health

A ten gram spirulina serving (20 tablets, or 1/3 ounce) supplies a rich profile of vitamins we need.

2.6. Spirulina vitamin content

Vitamins [a]	per 10 grams		U.S. DV		% DV
Vitamin A (beta carotene)	23000	IU	5000	IU	460 %
Vitamin C	0	mg	60	mg	0 %
Vitamin D	1200	IU	400	IU	300 %
Vitamin E (a-tocopherol)	1.0	IU	30	IU	3 %
Vitamin K	200	mcg	80	mcg	250 %
Vitamin Bl (thiamin)	0.35	mg	1.5	mg	23 %
Vitamin B2 (riboflavin)	0.40	mg	1.7	mg	23 %
Vitamin B3 (niacin)	1.40	mg	20	mg	7 %
Vitamin B6 (pyridoxine)	80	mcg	2	mg	4 %
Folate (folic acid)	1	mcg	0.4	mg	0 %
Vitamin B12 (cyanocobalamin)	20	mcg	6	mcg	330 %
Biotin	0.5	mcg	0.3	mg	0 %
Panthothenic Acid	10	mcg	10	mg	1 %
Inositol	6.4	mg	***		***

a. Earthrise Farms 1995.

Natural beta carotene (provitamin A)

Spirulina is the richest food in beta carotene, ten times more concentrated than carrots. Ten grams provide a remarkable 23,000 IU (14 mg) of beta carotene, 460% of the U.S. Daily Value (DV) of Vitamin A. High doses of Vitamin A may be toxic, but beta carotene in spirulina and vegetables is safe, because human bodies convert beta carotene to Vitamin A only as needed. Vitamin A is important in maintaining mucous membranes and pigments necessary for vision. Vitamin A deficiency is one of the most serious malnutrition diseases in the developing world, leading to blindness.

Beta carotene has therapeutic effects, including reducing serum cholesterol and ever present cancer risks. Over the past 12 years, cancer health authorities have published dozens of studies showing beta carotene reduces the risks of all kinds of cancers, including lung, throat, stomach, colon, gastrointestinal tract, breast and cervix.

Since the increase in cancer rates seems to be caused by environmental factors, especially diet, scientists say these risks can be reduced by increasing protective factors, especially beta carotene, in the diet. Natural beta carotene is a combination of *cis* and *trans* isomers, whereas synthetic beta carotene has only the *trans* form, with lower antioxidant potential.

2.7.	**Best beta carotene vegetables**[a]	
Food	serving size	IU of beta carotene
spirulina[b]	**1 heaping tbsp.** (10 g)	**23000**
papaya	1/2 medium	8867
sweet potato	1/2 cup, cooked	8500
collard greens	1/2 cup, cooked	7917
carrots	1/2 cup, cooked	7250
chard	1/2 cup, cooked	6042
beet greens	1/2 cup, cooked	6042
spinach	1/2 cup, cooked	6000
cantaloupe	1/4 medium	5667
chlorella[c]	**50 tablets** (10 g)	**5000**
broccoli	1/2 cup, cooked	3229
butternut squash	1/2 cup, cooked	1333
watermelon	1 cup	1173
peach	1 large	1042
apricot	1 medium	892

a. Vegetarian Times, "Recipes with A+ Nutrition", May 1986, pg 47.
b. Earthrise Farms, 1995.
c. Yaeyama Chlorella, 1995.

Although beta carotene is best known, spirulina contains an antioxidant rich complex of at least ten carotenoids. These mixed carotenes and xanthophylls function at different sites in the body and work synergistically with the other essential vitamins, Vitamin E, minerals and phytonutrients in spirulina. This is more effective than an isolated, synthetic beta carotene supplement.

Even if you don't eat the recommended 4 to 9 servings of fruits and vegetables every day (most people eat only 1 to 2 including french fries), get your natural carotenoid antioxidant protection from spirulina tablets or powder every day.

Vitamin B-12 and B-complex vitamins

Spirulina is the richest source of B-12, higher than beef liver, chlorella or sea vegetables. B-12 is necessary for development of red blood cells, especially in the bone marrow and nervous system. Although primary B-12 deficiencies, pernicious anemia and nerve degeneration, are quite rare, because B-12 is the most difficult vitamin to get from plant sources, vegetarians have taken to spirulina.

Ten grams contain 20 mcg of Vitamin B-12, 330% DV, using the approved microbiological assay. Vitamin B-12 label content claims for foods and dietary supplements are based on the approved microbiological assay. This method is used for spirulina, because it is being compared with the B-12 content of other foods and vitamins.

An alternative method developed in the 1980s, radioassay, has measured the B-12 assumed to be bioavailable to humans. Radioassay found higher levels of B-12 analogs and lower levels of bioavailable B-12 in all foods and supplements, and shows spirulina has only 20% of the original B-12. Even using these lower levels, it is the best non-animal source of Vitamin B-12.

Some incomplete research has suggested B-12 analogs could block B-12 absorption, based on limited results with very few individuals, and did not consider B-12 non-absorption due to folic acid or other dietary deficiency. In nearly 20 years, there have been no complaints of a vitamin B-12 deficiency from spirulina consumers, including children and vegetarians.

One tablespoon provides significant quantities of thiamin (23% DV), required for functioning of nerve tissues, riboflavin (23% DV), needed to gain energy from carbohydrates and proteins, and niacin (7% DV) needed for healthy tissue cells. Spirulina is a richer source of these vitamins than common whole grains, fruits and vegetables and some seeds.

Other B vitamins, B-6, niacin, biotin, panthothenic acid, folic acid, inositol and Vitamin E are also present in smaller amounts.

Naturally colloidal minerals

Algae absorbs many trace elements while growing and these minerals are well assimilated by the human body. Its mineral content varies depending on where it is grown and the minerals in the water.

2.8. Spirulina mineral content

Minerals[a]	per 10 grams	U.S. DV	% U.S. DV
Calcium	70 mg	1000 mg	7 %
Iron	10 mg	18 mg	55 %
Phosphorus	80 mg	1000 mg	8 %
Magnesium	40 mg	400 mg	10 %
Zinc	300 mcg	15 mg	2 %
Selenium	10 mcg	70 mcg	14 %
Copper	120 mcg	2 mg	6 %
Manganese	500 mcg	2 mg	25 %
Chromium	25 mcg	120 mcg	21 %
Sodium	90 mg	2400 mg	4 %
Potassium	140 mg	3500 mg	4 %
Germanium	60 mcg	-	-

a. Earthrise Farms, 1995.

The best natural iron supplement

Iron is the most common mineral deficiency worldwide, especially for women, children and older people. Women on weight loss diets typically do not get enough iron, and can become anemic. Iron is essential for strong red blood cells and a healthy immune system. Spirulina is a rich iron food, 10 times higher than common iron foods. Ten grams supply up to 10 mg of iron, 55% of the Recommended Daily Value.

Spirulina iron is easily absorbed by the human body. It is theorized that its blue pigment, phycocyanin, forms soluble complexes with iron and other minerals during digestion making iron more bioavailable. Hence, iron in spirulina is over twice as absorbable as the form of iron found in vegetables and most meats.[6]

Typical iron supplements are not well absorbed. Studies show iron in spirulina is 60% better absorbed than iron supplements such as iron sulfate. For this research refer to Chapter 4. For all people who need iron supplements, spirulina is one of the best sources.

2.9.	**Best food sources of Iron**[a]	
Food	serving size	mg Iron
Spirulina[b]	**1 tbsp. (10g)**	**10.0**
Chlorella[c]	**1 tbsp. (10g)**	**10.0**
Chicken liver, cooked	3 ounces	7.2
Crab, pieces, steamed	1/2 cup	6.0
Beef liver, fried	1/2 cup	5.3
Soybeans, boiled	1/2 cup	4.4
Blackstrap molasses	1 tbsp.	3.2
Spinach, cooked	1/2 cup	3.2
Beef, sirloin, broiled	3 ounces	2.9
Potato, baked	one	2.8
Scallops, steamed	3 ounces	2.5
Pistachios, dried	1/4 cup	2.2
Broccoli, cooked	1 spear	2.1
Cashews, dry-roasted	1/4 cup	2.1
Turkey, dark meat	3 ounces	2.0
Spinach, raw chopped	1/2 cup	0.8

a. The Complete Book of Vitamins and Minerals for Health, pg. 182.
b. Earthrise Farms, 1995.   c. Yaeyama Chlorella, 1995.

Calcium, magnesium, zinc and trace minerals

Spirulina is a concentrated calcium food, supplying more, gram for gram, than milk. Ten grams supply 7% DV for calcium. Calcium is important for bones and neural transmissions to the muscles. Deficiencies can lead to osteoporosis in older women. Ten grams supply 10% DV for magnesium, one of the most concentrated magnesium foods. Magnesium facilitates absorption of calcium and helps regulate blood pressure. Spirulina is low in iodine and sodium, and is no problem for those on salt-restricted diets.

Humans need dozens of essential trace minerals for the functioning of enzyme systems and many other physiological functions. Deficiency of trace minerals in the typical diet are thought to be widespread. Ten grams supply manganese (25% DV), chromium (21% DV), selenium (14% DV), copper (6% DV) and zinc (2% DV).

Essential fatty acids

Humans require a dietary source of essential fatty acids (EFA). They promote cholesterol normalization and are precursors for hormones, called prostaglandins. Spirulina has 4 to 7% lipids, or fats, and most of these are essential fatty acids. Ten grams have 225 mg of EFA in the form of linoleic and gamma-linolenic acid (GLA). The DV for an adult is a minimum EFA intake of 1% of total calories. Ten grams provide 8 to 14% DV, depending on sex and age group.[7]

GLA is the precursor to the body's prostaglandins – master hormones that control many functions. Dietary saturated fats and alcohol can cause in GLA deficiency and suppressed prostaglandin formation. Studies show GLA deficiency figures in many diseases and health problems, so a food source of GLA can be important.

2.10.	**Spirulina essential fatty acids**[a]	
	mg per 10 grams	**% total**
C 14:0 Myristic	1 mg	0.2 %
C 16:0 Palmitic	244 mg	45.0 %
C 16:1 Palmitoleic	33 mg	5.6 %
C 17:0 Heptadecanoic	2 mg	0.3 %
C 18:0 Stearic	8 mg	1.4 %
C 18:1 Oleic	12 mg	2.2 %
C 18:2 Linoleic	97 mg	17.9 %
C 18:3 Gamma-linolenic	**135 mg**	**24.9 %**
C 20 Others	14 mg	2.5 %
Total	**546 mg**	**100 %**

a. Earthrise Farms 1995.

The only other known sources of dietary GLA are mother's milk and oil extracts of evening primrose, black currant and borage seeds. Spirulina is a concentrated source of GLA, and a 10 gram serving has 135 mg. As a comparison, a daily dose of 500 mg of evening primrose oil has 45 mg. GLA comprises about 20 to 25% of the lipid fraction of spirulina, compared to only 9% for evening primrose oil.

2.11.	**Dietary sources of GLA**
Food sources	**Oil extracts**
Mother's milk	Evening primrose plant
Spirulina	Black currant and borage seeds

Phytonutrients

These functional nutrients have no published Recommended Daily Value, but are known to benefit health. They include glyco-lipids, polysaccharides, pigments and other growth factors.

A rainbow of natural pigments

Pigments help synthesize many enzymes necessary for regulating the body's metabolism. Spirulina's very dark color comes from these natural pigments which harvest different wave lengths of sunlight.

Phycocyanin ('algae-blue')

The most important pigment in spirulina, this protein complex is about 14% of the entire weight. Phycocyanin evolved a billion years before chlorophyll and may be the precursor to chlorophyll and hemoglobin. It has both magnesium and iron in its molecular formation, and therefore, phycocyanin may be the origin of life common to both plants and animals.[12] Research shown in Chapter 4 suggests it stimulates the immune system.

Chlorophyll (nature's green magic)

The common feature of green foods is their high chlorophyll content. Chlorophyll is known as a cleansing and detoxifying phytonutrient. Sometimes called 'green blood' because it looks like the hemoglobin molecule in human blood. Chlorophyll has a magnesium ion at its core, giving it a green color, and hemoglobin has iron, giving it a red color.[11] Spirulina's beneficial effect on anemia could be due to this similarity of chlorophyll and hemoglobin and its high bioavailable iron. Spirulina has 1% chlorophyll, one of nature's highest levels, and has the highest chlorophyll-a level. Chlorella has 2 to 3%, mostly chlorophyll-b.

Carotenoids (natural antioxidants)

About half of these yellow/orange pigments in spirulina are carotenes: Alpha, Beta and Gamma. About half are xanthophylls: Myxoxanthophyll, Zeaxanthin, Cryptoxanthin, Echinenone, Fuco-xanthin, Violaxanthin and Astaxanthin. Total mixed carotenoids make up 0.37% of spirulina. Although beta carotene is best known, this mixed carotenoid complex functions at different sites in the body and works synergistically to enhance antioxidant protection.

2.12.	**Spirulina natural pigments**		
Pigments[a]	**Color**	**per 10 grams**	**% total**
Phycocyanin	**(blue)**	1400 mg	**14 %**
Chlorophyll	**(green)**	100 mg	**1.0 %**
Carotenoids	**(orange)**	37 mg	**0.37 %**
Carotenes	*54 %*	*20 mg*	*0.20 %*
Beta carotene	45 %	17 mg	0.17 %
Other Carotenes	9 %	3 mg	0.03 %
Xanthophylls	*46 %*	*17 mg*	*0.17 %*
Myxoxanthophyll	19 %	7 mg	0.07 %
Zeaxanthin	16 %	6 mg	0.06 %
Cryptoxanthin	3 %	1 mg	0.01 %
Echinenone	2 %	1 mg	0.01 %
Other Xanthophylls	6 %	2 mg	0.02 %

a. Earthrise Farms 1995.

Polysaccharides

Spirulina contains only 15 to 25% carbohydrate and sugar. The primary forms of carbohydrates are rhamnose and glycogen, two polysaccharides which are easily absorbed by the body with minimum insulin intervention. Spirulina offers quick energy, without taxing the pancreas or precipitating hypoglycemia.[10]

Glycolipids and Sulfolipids

When the NCI announced that sulfolipids in blue-green algae were 'remarkably active' against the AIDS virus, attention was focused on the sulfolipid containing glycolipids (see Chapter 4). The three classes of lipids in spirulina are called neutral lipids, glycolipids and phospholipids. Glycolipids are 40% of the lipids, and contain sulfolipids. Sulfolipids in spirulina range from 2-5% of the total lipids.[8,9]

Enzymes

Enzymes are catalysts for chemical changes. There are thousands of enzymes, each catalyzing specific reactions. Dried spirulina contains a number of enzymes. One is superoxide dismutase (SOD), important in quenching free radicals and in retarding aging. SOD enzyme activity ranging from 10,000 to 37,500 units per ten grams has been found in spirulina powder from Earthrise Farms.

Comparing the green superfoods

Today more people understand the need for green vegetables than 20 years ago. Even fast food restaurants have installed salad bars. At the same time, there is growing concern with the quality of foods and vegetables grown on mineral depleted soils.

Green superfoods go beyond green vegetables because they are packed with beneficial nutrients. They go beyond isolated vitamin and mineral supplements, because as whole foods they are rich in functional nutrients and phytonutrients. Research reports link the phytonutrient, antioxidant and protective substances in plant foods with the prevention of degenerative diseases. This publicity has stimulated the greening of supplements with green superfoods.

Nutrient dense green superfoods are often consumed as tablets or capsules or by mixing powder in drinks. These ideal fast foods pick up your energy level, especially if you do not have time to eat the recommended 4 to 9 servings of fresh fruits and vegetables every day.

Green superfood supplements have become increasingly popular in the last 15 years. Spirulina and chlorella are specially cultivated algae. Aphanizomenon flos-aquae (referred to as 'blue-green algae') is harvested from a lake in Oregon. Barley grass and wheat grass are two superfoods from cultivated young cereal plants, harvested before they become grains. All five chlorophyll-rich foods are specially harvested to maximize purity, potency and quality.

How do they compare? Comparing price, aphanizomenon blue-green algae is the most expensive, followed by chlorella. Spirulina powder and tablets are less expensive, and barley grass and wheat grass powder and tablets are the least costly. Comparing nutrients, data provided by manufacturers is shown on the next page for a 10 gram serving.

Algae and grasses are the foundation of life on Earth, harvesting sunlight. Their deep green color glows with the vitality from the rainbow of natural pigments which power, protect and cleanse them while they grow. These natural foods will nourish, energize and cleanse your body naturally. Eating just a little of these concentrated green foods every day will benefit your health. Eating lower on the food chain will benefit the health of our planet.

2.13.   Green superfood nutrient comparison

Composition	spirulina[a] algae	chlorella[b] algae	aphaniz.[c] algae	barley[d] grass	wheat[d] grass
Protein	62 %	60 %	58 %	25 %	25 %
Carbohydrates	19 %	18 %	25 %	54 %	54 %
Fats (lipids)	5 %	10 %	5 %	4 %	4 %
Minerals (ash)	9 %	7 %	7 %	12 %	12 %
Moisture	5 %	5 %	5 %	5 %	5 %
Vitamins (per 10 grams)					
Beta carotene	23000 IU	5000 IU	12000 IU	5000 IU	5000 IU
Vitamin C	0 mg	4 mg	6 mg	31 mg	31 mg
Vitamin E	1 IU	1.5 IU	1.3 IU	3 IU	3 IU
Thiamin, B1	0.35 mg	0.17 mg	0.05 mg	0.03 mg	0.03 mg
Riboflavin, B2	0.40 mg	0.50 mg	0.50 mg	0.20 mg	0.20 mg
Niacin, B3	1.40 mg	2.80 mg	1.30 mg	0.75 mg	0.75 mg
Vitamin B6	80 mcg	140 mcg	110 mcg	128 mcg	128 mcg
Vitamin B12	20 mcg	5 mcg	32 mcg	3 mcg	3 mcg
Folacin	1 mcg	*	10 mcg	108 mcg	108 mcg
Biotin	0.5 mcg	*	3 mcg	11 mcg	11 mcg
Pantothenic acid	10 mcg	*	60 mcg	240 mcg	240 mcg
Inositol	6 mg	*	*	*	*
Minerals (per 10 grams)					
Calcium	70 mg	30 mg	140 mg	52 mg	52 mg
Iron	10 mg	10 mg	3.5 mg	6 mg	6 mg
Magnesium	40 mg	30 mg	22 mg	10 mg	10 mg
Sodium	90 mg	36 mg	27 mg	3 mg	3 mg
Potassium	140 mg	80 mg	120 mg	320 mg	320 mg
Phosphorus	90 mg	90 mg	50 mg	52 mg	52 mg
Zinc	0.3 mg	1.2 mg	0.2 mg	0.5 mg	0.5 mg
Manganese	0.5 mg	*	0.3 mg	1.0 mg	1.0 mg
Copper	120 mcg	*	40 mcg	200 mcg	200 mcg
Chromium	25 mcg	*	5 mcg	*	*
Phytonutrients (per 10 grams)					
Phycocyanin	1400 mg	none	*	none	none
Chlorophyll	100 mg	280 mg	200 mg	55 mg	55 mg
Total Carotenoids	37 mg	*	*	*	*
Gamma Linolenic Acid	135 mg	*	*	none	none
Glycolipids	200 mg	*	*	*	*
Sulfolipids	10 mg	*	*	*	*

a. Earthrise Farms, 1995.   b. Yaeyama Chlorella, 1995.
c. Cell Tech, Alpha Sun.   d. Cereal Grass, ed. by Ronald Seibold.   * no data available.

Chapter 3

Self-care programs with clean green energy

Spirulina is most effective when used in a natural food diet as one part of a personal strategy for self-care. This strategy embraces recommendations by the National Academy of Science, the National Cancer Institute and the American Heart Association. They call for more whole grains and vegetables and less fat, salt and sugar.

Following these guidelines will lower the risks of cancer, heart and degenerative diseases.

- Eat lighter as part of a natural weight control plan.
- Vegetable protein and B-12 for vegetarians.
- Reducing cholesterol and PMS.
- Ideal for fasting and for body cleansing programs.
- Energy and endurance for athletes and body builders.
- Easy-to-digest for older people on restricted diets.
- Anti-aging strategy.
- Great for children and mothers.
- The best storage and survival food.

How to use spirulina powder

100% pure powder is a uniformly dark green or blue-green color and has no other colored particles. Your body feels energy within minutes because the powder is naturally digestible. It provides quick energy and nourishment between meals or in place of a meal. Some have asked whether you can take too much. It is a perfectly safe natural food. Some people take two tablespoons or more each day.

3.2. Author with a morning smoothie in a blender.

The most popular way to enjoy it at home is to add it to your favorite fruit or vegetable juice in a blender. Start with one teaspoon (5 grams) and add flavors or spices to suit your taste. Later on you can increase the amount. Many regular users take one heaping tablespoon (10 grams) per drink. Try out our successful green smoothie recipes!

Hand held micro mixers are handy for mixing powder directly in a glass of water or juice. Easier to clean up than a big blender. Micro mixers use two batteries, and you can carry one with you in a purse or briefcase. Now you can take powder on the road, use it at work, or even in a hotel room. No longer do these circumstances keep you from your healthy green drink.

Helpful hints: Don't stick a wet spoon in your spirulina bottle or put a spoonful directly into liquid. Water will stick the powder on your spoon. Add it to liquid slowly while stirring. Spirulina will keep well if handled properly. The dry powder absorbs water from the air if you leave it open, so keep the bottle tightly sealed when not in use. You don't need to refrigerate it, but do keep in a cool, dry, dark place.

Add a teaspoon to a tablespoon to dishes to enhance nutritional content. It's tasty in soups, salads, pasta, breads and taboulie. Even a little will give food a dark green color. With recipes that require cooking, heat spirulina as little as possible. Like many natural foods, heat can damage its sensitive nutrients. Creative recipes are found in Spirulina, *The Whole Food Revolution* by Larry Switzer[1] and *The Spirulina Cookbook* by Sonia Beasley.[2]

3.3. Delicious green energy drinks!

Morning Smoothie
An instant breakfast to start your day

Blend 1 tablespoon spirulina in 2 cups of tropical blend juice (or orange, apple or pineapple juice). For options you can add one whole fruit (banana, orange or peach), almonds, sunflower seeds or even flavors such as vanilla or lime to suit your taste. (Mix well in a blender. Makes 2 servings).

Veggie Cocktail
A mid afternoon vegetable pick up

Blend one tablespoon spirulina in 2 cups of vegetable juice (or carrot or tomato juice). For options you can add whole vegetables, herbs (parsley, dill weed) or spices (cayenne, horseradish) to suit your taste. (Mix well in a blender. Makes 2 servings).

Convenient tablets and capsules

Quality tablets can be made without sugar, starch, fillers, animal parts, preservatives, stabilizers, colors, coatings, and with only a minimum of vegetable tableting agents. Made in this way, the color of the tablet should be a uniform dark green without light colored spots or specks. Capsules should also be free of excess fillers or additives, and now, vegetable capsules are available.

Many bottles provide nutritional information for a six tablet serving (3 grams), but you can take more if you like. Often, people take 10, 15 or 20 tablets or capsules a day. Twenty 500 mg tablets are equal to a heaping tablespoon of powder (10 grams).

Tablets deliver the same benefits as powder, but digestion takes about an hour. For faster results, some people chew or dissolve the tablets in the mouth. Because it is a natural whole food, you can take tablets by themselves between meals.

If you are using spirulina to balance your diet and help eat lighter meals, take tablets or capsules an hour before you eat. If there is a time of day when your energy runs low, take some tablets and see how your body feels one or two hours later. Both tablets and capsules are helpful with water after coffee or alcohol.

How people are using this superfood

The following self-care programs show how people use spirulina to transform their health and vitality. It's best to examine your dietary habits and eliminate high fat, high meat, high sugar and junk food diets. You will notice spirulina is more effective as part of a lighter, fresher, more natural diet. These recommendations follow dietary programs advocated by leading medical experts for reducing weight, cholesterol and pre-menstrual stress, and for enjoying a longer, healthier life.

The problem with today's modern diet

Our modern diet is driven by the appetites generated by consumer advertising. Many people are sold on fast convenience foods – usually rich in fats, carbohydrates and sugar, and low in natural vegetables and fiber. These foods typically increase body weight, raise cholesterol levels, and worsen digestive and colon problems in later years. We often eat food in the wrong combinations, or out of compulsion or nervousness, and do not heed our true appetite signals.

Many conventionally grown and over-processed foods loaded with chemicals have low nutritional value, especially low in essential trace elements. The human digestive system, when overloaded with fatty non-nutritious foods, doesn't assimilate enough quality nutrients. Under these conditions, the body is continually starving for more nutrients, triggering appetite and compulsive overeating. Without exercise, extra calories stay on as fat. Overweight people often remain trapped in this cycle, a difficult one to break.

Crash diets, fad diets and drug diets do not work

Sudden weight loss with diet drugs or crash diets is very stressful and may have adverse side effects. Research shows common diet drugs containing phenylpropanalomine (PPA) may be addicting and have dangerous side effects on the kidneys and heart. Weight loss from drug diets is often temporary, and when the weight returns, the percentage of fat to body weight is higher than before dieting. This vicious cycle continues because the real problem is neglected – improper diet and its accompanying reinforcing attitudes.

Natural weight control

A more natural diet satisfies hunger because it satisfies the body's real hunger for nutrition. Spirulina is a very concentrated natural food. As part of a wholesome natural food diet suggested in the following pages, it can help restore natural body weight. Many people use it along with a low carbohydrate diet and exercise to eat lighter meals and avoid fattening snacks.

Take a heaping teaspoon of powder (about 5 grams), or at least 6 tablets one hour before meals or snack breaks. You know when you're going to be hungry, so plan ahead. Tablets take a little more time to be assimilated than the powder.

This green superfood can help satisfy a body's appetite. It is not an appetite suppressant in any way, and contains no drugs or chemicals that trick the body. It is simply concentrated, easily digested natural nutrition. Especially important to dieters, it is rich in iron, often found deficient in women on low calorie diets. One resource book written on this subject is *The Spirulina Diet* by Dr. Saundra Howard.[3]

At mealtimes, eat a balanced diet of natural foods, minimizing high calorie fat foods. Exercise daily to burn off calories and fat deposits, and to maintain a toned body. Your goal should be to lose weight slowly while you are reprogramming your eating habits. Both must happen together for long term success. This way you can graduate from the destructive cycles of food binging, crash diets and diet drugs.

Because metabolism and biochemistry are different for each person, weight loss results may differ. Track your weight over several weeks. If you want to strengthen your program, increase the amount of spirulina slowly in order to eat lighter meals. It is important to eat regular nutritious meals.

Slow but steady weight loss is desirable. Often dieters are able to stabilize their body weight at a more ideal lower level. Spirulina can help us remember the wisdom of a natural diet. Most of all, it helps us to lighten up, and provides the energy to make the switch from a bulky unhealthy diet to lighter, more powerful nutrient rich foods.

Emphasize a natural food diet

Here are dietary suggestions to eat lighter, low-calorie foods, get plenty of exercise, and supplement your diet with spirulina for extra nutritional support.

Switch to natural foods

Add more of these fresh, natural, high fiber, low fat foods to your diet. They provide the nutrition you need. Prepare foods simply, without adding sugar and with a minimum of oil, butter and salt. Natural herbs and spices can enhance the flavor.

Fresh fruits, vegetables and legumes contain natural vitamins and minerals. Canned or frozen foods are often loaded with sugar, salt and preservatives.

Whole grain breads and cereals provide excellent nutrition and a source of natural fiber, important for digestion and regularity. Once you use whole grain foods, you may find white bread, polished rice and sugary cereals less appealing.

Low-fat milk, yogurt, and cheeses have lower calories. Add fresh fruit to plain yogurt and cottage cheese, and avoid the excess sugar in flavored products. Use naturally aged cheese rather than processed cheese spreads.

Fresh fish and fowl are good sources of lean protein. Look for chicken and turkey that is not grown on factory farms, where birds are pumped with antibiotics and drugs to prolong their lives in unhealthy conditions. Avoid toxic foods.

Juices, herbal teas, mineral water and fresh water are healthy liquids. Use beverages made from natural ingredients free of added chemicals, flavors, colors, and sugars.

Exercise for more energy

Regular exercise is an important part of any weight loss program. Find several activities you enjoy regularly. It might be aerobics, tennis, jogging, swimming, bicycling or brisk walking. Drink lots of fresh clean water. Breathe deeply and oxygenate your entire body. Regular exercise will give you a more positive outlook towards life.

Three day suggested meal plan for a lighter, more natural diet

For a natural dietary program to lose weight, keep your total calories below 1100 calories per day with meals like those below. As much as possible use fresh, organically-grown produce to minimize intake of pesticides and herbicides.

Day One

Breakfast: 1 egg, 1/2 broiled tomato with basil, 1 slice whole wheat toast, 1 pat butter, grain coffee or herbal tea.

Lunch: *(6-10 spirulina tablets 1 hour before)* 1 small tossed green salad, 2 tbsp. lo-cal dressing, 4 oz. sauteed lean fresh fish, 1/2 cup fresh green beans, 1 tbsp. slivered almonds, herbal tea, mineral water.

Dinner: 2 lean pieces tarragon chicken, 1/2 cup steamed brown rice, 1/2 cup steamed broccoli, herbal tea.

Day Two

Breakfast: 1 cup oatmeal, 1/2 cup low-fat milk, 1 tbsp. raisins, 1 bran muffin, 1 pat butter, grain coffee or herbal tea.

Lunch: *(6-10 spirulina tablets 1 hour before)* 3 oz. water-packed tuna sandwich on whole wheat with alfalfa sprouts, lettuce and tomato, 1 cup homemade minestrone soup, herbal tea or 6 oz. low-fat milk.

Dinner: 1 green pepper stuffed with brown rice, 1 steamed ear of corn, 1/2 fresh apple, iced herbal tea, mineral water.

Day Three

Breakfast: 6 oz. orange juice, 1 cup bran cereal, 1/2 cup low-fat milk, 1/2 sliced banana, wheat toast, 1 pat butter, grain coffee.

Lunch: 1 cup gazpacho soup, 1 cup taboulie salad, 4 cherry tomatoes, 4 cucumber slices, celery sticks, iced herb tea.

Dinner: *(6-10 spirulina tablets 1 hour before)* 1 spinach salad, 2 tbsp. lo-cal dressing, 4 oz. baked red snapper with lemon and parsley, 1/2 cup steamed zucchini, 2 crackers, mineral water.

Lowering cholesterol levels

Well-publicized medical studies show as cholesterol levels decline, risks of heart attacks and strokes decline as well. Over 70% of Americans now understand the need to reduce cholesterol-rich foods in their diet. These foods increase blood serum cholesterol which creates plaque and blocks the arteries, causing atherosclerosis leading to heart disease.

More people are testing their cholesterol levels. Total cholesterol level above 200 mg/dl is considered risky, and above 240 is considered high risk. Below 200 is considered acceptable, but ideal may be 150 to 160. For specific cholesterol guidelines, consult a physician or health practitioner. Excessive cholesterol may require drug therapy.

Spirulina is a helpful food for cholesterol reduction. Research in Japan with male volunteers shows only 4 grams of spirulina a day (about 8 tablets) significantly reduced cholesterol levels. Refer to Chapter 4 for details of this study.

In any event, says *Medical Self Care,* "The key to cholesterol reduction is diet. For most people, this involves only simple adjustments and solutions. Most people can lower cholesterol by 30 to 40 mg/dl within a few weeks by making basic dietary changes."[4] Changes include:

---

1. Reduce saturated fat by eating less red meat, whole fat dairy products, commercial products with lard, coconut and palm oils and fried foods. Eat more fish, poultry, low-fat milk and cheese.

2. Reduce dietary cholesterol by eating fewer eggs, less organ meat, animal and dairy products. Eat more vegetables and fruits.

3. Eat more soluble fiber such as pectin in fruits and vegetables and guar gum in oats.

4. Eat more fish because omega-3 fatty acids in cold water fish are known to lower cholesterol.

5. Read food labels and avoid hydrogenated oils.

6. Exercise.

7. Control weight to decrease the risk of heart attack.

Reducing pre-menstrual syndrome (PMS)

Once a month, millions of women experience the unpleasant and disruptive effects of their hormonal cycles. Studies show women with more severe PMS have unusually low levels of certain nutrients, so many health experts urge a nutritional approach.

Three key factors increase the severity of PMS – poor nutrition, lack of exercise and stress. By improving the quality of foods, eating less of certain foods, exercising regularly, and learning to reduce stress, women can feel better all month.

Five steps can reduce the severity of PMS. Results are cumulative. Many of these recommendations are the same as those in the program for weight control and lowering cholesterol.

---

1. Enjoy fresh, natural foods.
Eat more fresh fish and poultry, whole grains, nuts, fresh fruits, vegetables. Drink plenty of water, herb teas and fruit juices. Prepare food simply, using a minimum of oil, butter, and salt.

2. Avoid certain foods and drugs.
Cut down or avoid salt, sugar, white flour, convenience foods full of chemicals, dairy products and red meat, chocolate, soft drinks and coffee. Studies show women with PMS consume more of these foods than women who do not. Cigarettes and coffee can deplete your body of nutrients and aggravate PMS.

3. Supplement with key vitamins and minerals.
Many clinics recommend foods or supplements rich in B-complex, magnesium, zinc, beta carotene, GLA and other vitamins, minerals and herbs. By containing many of these nutrients, spirulina is useful in a PMS reducing plan, and several PMS supplements contain spirulina.

4. Exercise.
Vigorous activity will increase your blood flow, oxygenate tissues, and help lift depression and anxiety.

5. Discover the art of relaxation.
Stress makes PMS worse. Create a more relaxing environment. Learn to let go through muscle relaxation, talking to friends, or meditation.

Lowering risks of cancer

Scientific organizations such as the National Cancer Institute have published guidelines for lowering the risks of all kinds of cancers. Poor diet is associated with 35% of cancer deaths.[5] High-fat, low-fiber diets greatly increase risks of prostate, colon, gland and breast cancer.

Food recommendations, such as those mentioned already, include reducing fat consumption, reducing caloric intake to maintain desirable body weight, and eating lean meats, poultry, fish and low-fat dairy products, more fruits, vegetables, beans, peas and whole grain cereals and breads.

Tobacco use is associated with 35% of cancer deaths, and should be avoided. Excessive alcohol consumption is associated with 3% of cancer deaths. Together with diet, reducing drugs like tobacco and alcohol significantly reduces cancer risks. About 27% of cancer deaths are not directly related to diet or drugs, but result from environmental, occupational or stress factors.

Over the past two decades, well-publicized scientific studies have clearly shown that eating foods rich in beta carotene will lower the risks of all kinds of cancer. See a summary in Chapter 4.

One of the most famous reports was published by The National Research Council in 1982. *Diet, Nutrition and Cancer*[6] concluded that foods rich in beta carotene and Vitamin A reduce cancer risks. The study recommended two servings each day of vegetables rich in beta carotene such as carrots, sweet potatoes, green vegetables and squashes, and fruits such as papaya and cantaloupe. Unfortunately, many people do not eat these two vegetable servings each day, so over a dozen new studies are testing the beneficial effects of beta carotene supplements.

Fortunately, spirulina is over ten times more concentrated in beta carotene than any other vegetable. Only three grams daily offers 140% of the US RDA of Vitamin A in the safe form of natural beta carotene. Plus, its entire carotenoid complex offers antioxidant protection at different sites in the body.

Fasting and cleansing

Fasting one day a week is a common practice in traditional societies and in many religions. The purpose of going without solid food for three days to a week or longer is to allow the body to cleanse and renew itself. People who benefit from prolonged fasting report a feeling of detoxification which makes them feel physically stronger and psychologically clearer.[7]

One of the first discoveries was spirulina's value as an aid for fasting. It eases intestinal problems because it is easy to digest and provides energy and stamina for work and play during a fast. Because it is a light, low calorie food providing essential nutrients, and because it digests easily, spirulina makes fasting easier and more effective.

A well known booklet on this subject is *Rejuvenating the Body through Fasting with Spirulina Plankton* by Dr. Christopher Hills. He writes: "as a source of nutrition during fasting or dieting, it is excellent because it helps cleanse the intestinal tract as well as relax the smooth muscle of the bowels."[8]

"Fasting with spirulina and mixing it with fruit and vegetable juices is the perfect and most natural way to flush out the system with liquids and chlorophyll without denying the body the nutrients for full and effective metabolism."[9] Dr. Hills suggests a seven day fast with fruit, fruit juices, water and spirulina, without any other solid food.

Fasting requires discipline and should be done for limited periods. Fasting should be accompanied with common sense, and when it is over, light simple meals should be phased in slowly over several days before moving back to a normal diet.

The first few days of fasting may result in discomfort due to detoxification. If side effects become severe, fasting should be stopped. Fasting is not for everyone, and people with special dietary or medical problems should consult their health practitioner first.[10]

Colon cleansing

In the modern low-fiber diet, wastes (or mucoids) can accumulate in pockets in the colon. Never removed, these wastes can lead to constipation, weakened digestion, poor nutrient absorption, and in later life, colon cancer, one of the most common cancers today. These problems increase with age, as seniors will confirm.

A complete plan for health through colon rejuvenation is found in *The Colon Health Handbook*, by Robert Gray. In the book, he describes yeast and spirulina as the only protein supplements that are not mucoid-forming in the intestines.[11] Spirulina is described as a 'metabolic activator', which acts directly on the body tissues at the cellular level to promote increased activity to 'burn up' mucoforming substances.

Spirulina "is very energizing, generally superior to all ginsengs, dong quai, bee pollen and vitamin B-15. The combination of spirulina's energizing and appetite suppressant properties has made it popular for use while fasting. It is an aggressive cleansing herb that empties toxins out of the body tissues into the lymph."

Gray describes the value of lactobacillus. "A healthy population of lactobacillus within the intestinal tract far outstrips all metabolic activators known to the author.[12] In order for the counter-mucoid effect of dietary fiber to be significant, there first must be a good implantation of lactobacteria in the intestinal tract. The fiber then acts to keep the lactobacteria at a high level of activity and to avoid their extinction."[13]

Research in Japan has shown laboratory rats eating spirulina have higher levels of lactobacillus and B-vitamin absorption from the entire diet. For more on this study, see Chapter 4.

Many older people suffer from chronic constipation, necessitating laxatives for symptomatic relief. Others face the prospect of colon cancer. In Brazil, people have used spirulina to relieve constipation. Many high income Brazilians, who eat excess beef with few vegetables and fiber, complain of constipation and take spirulina for its cleansing and normalizing effect.

Athletes and bodybuilders

Athletes need extra nutrition. Because conventional foods no longer contain the nutrients they once had, concentrated superfoods have become popular. Taken before jogging or athletic competition, athletes say it delivers energy and improves stamina. It increases the endurance of marathon runners. Backpackers, cyclists and mountain climbers who carry all their food take tablets for more energy and stamina per weight than conventional foods.

Spirulina is a high intensity food perfectly suited for high intensity training. It contains GLA which is known to stimulate prostaglandins, master hormones which regulate every cell of the body, including heart, skin, circulation and musculature. Correct prostaglandin levels are necessary for good health and performance.[14]

Proteins are essential for proper endurance training, and are needed to regenerate body tissue, providing the framework for muscles, tendons and blood hemoglobin. For the bodybuilder, spirulina offers 65% protein, easy to digest and low in fat. Bodybuilders take 10 grams up to three times a day. Before competition spirulina gives an energy boost with sustaining power, mixed with milk, egg, honey and juice. Taken before meals it can satisfy appetite and help reduce caloric intake, essential for maintaining competitive weight.[15]

World Class and Olympic athletes in China and Cuba use spirulina to improve performance. At the largest Chinese training center for over 2000 athletes, trainers report it improves recovery for all athletes and boosts the immune system. This allows these athletes to intensify training, something for which they are renown.

The Cuban Ministry of Sports was given 1,600 bottles of spirulina before the 1996 Olympic games so athletes could intensify training. Cuban athletes, especially track stars, have consumed spirulina for many years. It helps create and mend muscle mass. In heavy training, fast burners have trouble retaining iron. Spirulina prevents anemia because it has ten times more iron than spinach. Some marathoners consume 8 grams daily for endurance and to ward off cramping. It helps eliminate carbon dioxide faster, so the muscles get oxygen faster. When training increases appetite, its nutrition helps curb hunger pangs.[16]

Children, pregnant and nursing mothers

Nutrition during pregnancy is especially important. Iron deficiency anemia in mothers and children is the most prevalent nutritional disorder.[17] Pregnant women need spirulina's extra easy-to-digest protein and bioavailable iron, without more saturated fats. In India and Vietnam, it's prescribed for pregnant and nursing mothers.[18]

Parents are often surprised at how their children enjoy spirulina. Children like to suck on tablets and many enjoy munching on spirulina covered popcorn. Often a jar of tablets may have to be hidden just like a jar of cookies. Children often display a big green smile.

3.4. Serena, USA.
(Courtesy B. Lackey)

3.5. Janette, Germany
(Courtesy Kunst & Kergen)

3.6. Yuki, Japan.
(Courtesy H. Shimamatsu)

3.7. Tess, England.
(Courtesy S. St. Clair-Ford)

A self-care program for seniors

Spirulina is a often thought of as a superfood for people in their 20s, 30s and 40s, to boost their high energy lifestyle. Not so in Japan, according to a 1988 customer survey.[19] It reported 73% of spirulina customers are 50 years and older, and 57% of these people are women. Why? 45% say they are taking it for treatment of a specific problem. Another 28% take it to maintain their health and restore physical wellness, and 12% use it as a nutritional supplement. Of those taking it for treatment, 22% use it for blood sugar problems and diabetes, 15% for eye problems, and 14% to relieve constipation.

Japanese seniors do not consider spirulina a short term fad. They are serious about maintaining long term health. and have lower medical bills than Americans. Typically, Japanese take 4 grams every day (equivalent to eight 500 mg tablets), and many use even more as part of their regular program for maintaining long term health.

Because many older people don't eat enough, have restricted diets, or suffer from poor digestion, many have low energy and may be undernourished. Some seniors eat by themselves and don't make the effort to have nutritionally sound meals. Concentrated foods like spirulina are an excellent supplement: 60% protein, easy-to-digest, and known to assist recovery from malnutrition. All vegetable low-fat protein means seniors can lighten up on meat centered diets that aggravate arthritis and raise cholesterol.

With the highest level of the protective anti-oxidant beta carotene, it's good for eyes and vision. It builds healthy lactobacillus, aiding in assimilation and elimination. More older people desire to eat less meat, so it's important to choose iron rich vegetable sources like spirulina. Vitamin B-12 absorption decreases with age, so B-12 rich spirulina is a sound choice. It contains the rare essential fatty acid GLA, essential nutrients for healthy skin. Gram-for-gram it has more calcium and magnesium than other foods. Calcium and magnesium supplements are recommended for mature women who may have lost calcium from their bone mass and suffer from osteoporosis.

People over 50 in particular are likely to notice the benefits of eating a little every day.

Anti-aging strategy

Americans are becoming more aware of the value of nutritional supplements for an aging population. *Business Week* (October 9, 1989)[20] identified a new category for the food industry: "Nutrition of the Future – Foods that Fight Aging," highlighting the role of antioxidant nutrients: vitamins C, E and beta carotene. "Antioxidants sop up rogue oxygen atoms, called free radicals, that some researchers believe roam through the bloodstream, wrecking havoc on cells and organs and increasing aging effects and the risks of cancer."

Anti-aging foods and supplements for older people who don't eat much, eat inappropriately, or can't absorb enough nutrients will probably be in the form of high-nutrient foods. This practically describes spirulina: a digestible whole food with the highest beta-carotene level of any food. Now big multi-vitamin companies are jumping on the bandwagon, promoting the addition of beta-carotene because of its anti-cancer and anti-aging effects. But they're adding synthetic beta carotene. Already, research has shown natural beta carotene from vegetables and algae like spirulina has far greater antioxidant properties than the synthetic.

Help recovery the morning after

Many international businessmen work hard all day, socialize with clients and co-workers at night, and the next day rise early, ready to go again. Some have discovered a new secret. Ten or more spirulina tablets with two glasses of water after an evening on the town can reduce or even prevent the worst effects of hangovers the next day.

After too much alcohol, two glasses of water are essential for rehydration. Spirulina adds protein, vitamins and minerals to rebuild the body's depleted nutrients. It does not aid sobriety, but with water, helps avoid that dehydrated and depleted feeling the next morning.

3.8. Japanese hangover relief supplement with spirulina.

Survive and thrive with superfood reserves

Millions of people store several years of food for security in an uncertain world. Their concerns are earthquakes, environmental catastrophe, poisoning of land and water, nuclear accidents, food supply interruption or economic collapse. Natural foods are becoming more popular in reserves. Spirulina has always been popular with people who store food because it is concentrated, lightweight and portable – perfect for an emergency food supply.

Algae can supplement an existing food reserve. Its extra nutrition can turn a storage food diet from merely surviving to really thriving. Much less space is needed compared to bulky grains, and it is ideal for urban or suburban dwellers with limited space. In hard economic times, if paper currency were to become devalued, spirulina would hold its value for trade. It may even be better than gold, which you can't eat. Even for those who become temporarily unemployed or suffer loss of income, a green stash can always come in handy.

For long term storage, keep it in airtight and watertight containers, without exposure to light or excessive heat. It is best to rotate these foods into the daily diet and replace with fresh stocks. Regular plastic jars may not protect against oxygen penetration which can destroy valuable beta carotene over time. Oxygen barrier plastic jars are available which will allow storage for several years. One company offers volume discounts. Cases of pound size jars allow daily use from one jar, while keeping the remaining spirulina safely stored and sealed.

3.9. Superfood storage jar.

Enjoy this green superfood every day

Regardless of your personal self-care strategy, people who take it each day report the most health benefits, whether it is for energy, building, cleansing, or dieting. One additional benefit you may notice from regular use is more beautiful skin. We need all the protection we can get from eating the right foods like spirulina.

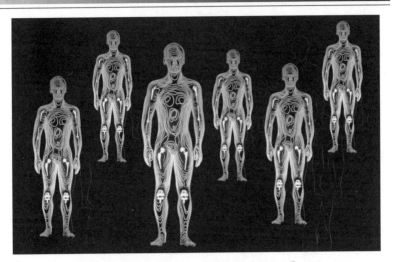

Chapter 4

New research reveals health benefits

An international detective hunt has been underway for 20 years. Researchers in Japan, China, India, Europe and the USA are discovering how and why this microalgae is so effective for human and animal health. Hundreds of published scientific studies reveal how spirulina, and its unique phytonutrients and extracts, boost the immune system and improve health in many ways.

- Increase anti-viral activity.
- Lower cholesterol.
- Reduce risk of cancer.
- Stimulate the immune system.
- Reduce kidney toxicity.
- Build healthy lactobacillus.
- Improve wound healing.
- Eliminate malnutrition.
- Reduce radiation sickness.

(This information is solely for education and information purposes. It is not intended as medical advice. People with medical questions should consult their physician or health professional.)

Latest Scientific Research

Effects on the AIDS Virus, Cancer and the Immune System.

by Ronald Henson and Richard Kozlenko DPM, Ph.D. M.P.H.

There are several new peer reviewed scientific studies about spirulina's ability to inhibit viral replication, strengthen both the cellular and humoral arms of the immune system and cause regression and inhibition of cancers. While these studies are preliminary and more research is needed, the results so far are exciting.

Potent Anti-Viral Activity

In April 1996, scientists from the Laboratory of Viral Pathogenesis, Dana-Farber Cancer Institute, Harvard Medical School and Earthrise Farms announced on-going research, saying *"Water extract of Spirulina platensis inhibits HIV-1 replication in human derived T-cell lines and in human peripheral blood mononuclear cells. A concentration of 5-10 µg/ml was found to reduce viral production."*[1]

HIV-1 is the AIDS virus. Small amounts of spirulina extract reduced viral replication while higher concentrations totally stopped its reproduction. Importantly, with a therapeutic index of >100, spirulina extract was non-toxic to the human cells at concentrations stopping viral replication.

Another group of medical scientists in Japan has published new studies regarding a purified water extract unique to spirulina named *Calcium-Spirulan*. It inhibits replication of HIV-1, Herpes Simplex, Human Cytomegalovirus, Influenza A virus, Mumps virus and Measles virus in-vitro yet is very safe for human cells. It protects human and monkey cells from viral infection in cell culture. According to peer reviewed scientific journal reports this extract, *"holds great promise for treatment of HIV-1, HSV-1, and HCM infections, which is particularly advantageous for AIDS patients who are prone to these life-threatening infections."*[2]

Calcium-Spirulan is a polymerized sugar molecule unique to spirulina containing both Sulfur and Calcium. Hamsters treated with this water soluble extract had better recovery rates when infected with an otherwise lethal Herpes virus.[3]

How does it work? When attacking a cell, a virus first attaches itself to the cell membrane. However, because of spirulina extract, the virus cannot penetrate the cell membrane to infect the cell. The virus is stuck, unable to replicate. It is eventually eliminated by the body's natural defenses. Spirulina extracts may become useful therapeutics that could help AIDS patients lead longer more normal lives.

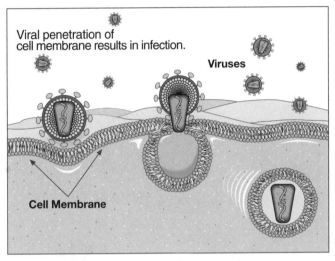

Viral penetration of cell membrane results in infection.

Viruses

Cell Membrane

4.2. Spirulina prevents viral penetration of the cell membrane.

Anti-Cancer Effects

Several studies show spirulina or its extracts can prevent or inhibit cancers in humans and animals. Some common forms of cancer are thought to be a result of damaged cell DNA running amok, causing uncontrolled cell growth. Cellular biologists have defined a system of special enzymes called Endonuclease which repair damaged DNA to keep cells alive and healthy. When these enzymes are deactivated by radiation or toxins, errors in DNA go unrepaired and, cancer may develop. In vitro studies suggest the unique polysaccharides of spirulina enhance cell nucleus enzyme activity and DNA repair synthesis. This may be why several scientific studies, observing human tobacco users and experimental cancers in animals, report high levels of suppression of several important types of cancer. The subjects were fed either whole spirulina or treated with its water extracts. [20,27,28]

Strengthens Immune System

Spirulina is a powerful tonic for the immune system. In scientific studies of mice, hamsters, chickens, turkeys, cats and fish, it consistently improves immune system function. Medical scientists find it not only stimulates the immune system, it actually enhances the body's ability to generate new blood cells. Important parts of the immune system, Bone Marrow Stem Cells, Macrophages, T-cells and Natural Killer cells, exhibit enhanced activity. Spleen and Thymus glands show enhanced function. Scientists also observe spirulina causing macrophages to increase in number, become "activated" and, more effective at killing germs.

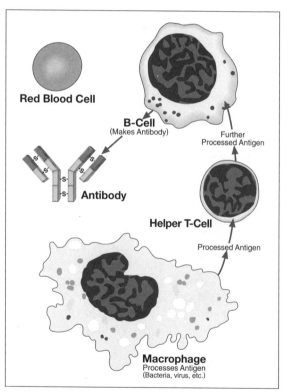

4.3. Key players in immunity that are stimulated in the presence of spirulina or its extracts.

Feeding studies show even small amounts build up both humoral and cellular arms of the immune system.[30] Spirulina accelerates production of the humoral system (antibodies and cytokines), allowing it to better protect against invading germs. The cellular immune system includes T-cells, Macrophages, B-cells and the anti-cancer Natural Killer cells. These cells circulate in the blood and are especially rich in body organs like the liver, spleen, thymus, lymph nodes, adenoids, tonsils and bone marrow. Spirulina up-regulates these key cells and organs, improving their ability to function in spite of stresses from environmental toxins and infectious agents.[4,25,28,29,30,31]

Phycocyanin Builds Blood

Spirulina has a dark blue-green color, because it is rich in a brilliant blue polypeptide called Phycocyanin. Studies show it affects the stem cells found in bone marrow. Stem cells are "Grandmother" to both the white blood cells that make up the cellular immune system and red blood cells that oxygenate the body. Chinese scientists document Phycocyanin stimulating hematopoiesis, (the creation of blood), emulating the affect of the hormone erythropoetin, (EPO). EPO is produced by healthy kidneys and regulates bone marrow stem cell production of red blood cells. Chinese scientists claim Phycocyanin also regulates production of white blood cells, even when bone marrow stem cells are damaged by toxic chemicals or radiation.[23]

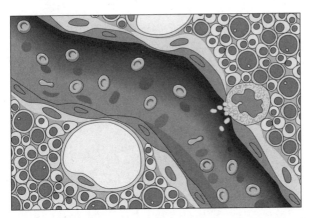

4.4. Spirulina or its extracts have demonstrable effect in stimulating production of new red and white blood cells.

Based on this effect, spirulina is approved in Russia as a "medicine food" for treating radiation sickness. The Children of Chernobyl suffer radiation poisoning from eating food grown on radioactive soil. Their bone marrow is damaged, rendering them immunodeficient. Radiation damaged bone marrow cannot produce normal red or white blood cells. The children are anemic and suffer from terrible allergic reactions. Children fed just five grams in tablets each day made dramatic recoveries within six weeks.[75]

Other Potential Health Benefits

Spirulina is one of the most concentrated natural sources of nutrition known. It contains all the essential amino acids, rich in chlorophyll, beta-carotene and its co-factors, and other natural phytonutrients. This is the only green food rich in GLA essential fatty acid. GLA stimulates growth in some animals and makes skin and hair shiny and soft yet more durable. GLA also acts as an anti-inflammatory, sometimes alleviating symptoms of arthritic conditions.

Spirulina acts as a functional food, feeding beneficial intestinal flora, especially Lactobacillus and Bifidus. Maintaining a healthy population of these bacteria in the intestine reduces potential problems from opportunistic pathogens like E. coli and Candida albicans. Studies show when spirulina is added to the diet, beneficial intestinal flora increase.

Conclusion

Based on this preliminary research, scientists hope the use of spirulina and its extracts may reduce or prevent cancers and viral diseases. Bacterial or parasitic infections may be prevented or respond better to treatment and wound healing may improve. Symptoms of anemia, poisoning and immunodeficiency may be alleviated.

Scientists in the USA, Japan, China, Russia, India and other countries are studying this remarkable food to unlock its potential. More research is needed to determine its usefulness against AIDS and other killer diseases. However, it is already clear this safe and natural food provides concentrated nutritional support for optimum health and wellness.

Cholesterol reduction

By now, Americans are well aware of the need to lower cholesterol levels in order to lower the risks of heart attacks and strokes, the number one cause of death. Besides dietary improvements, the search is underway to identify natural foods having a cholesterol reducing effect, such as fish oil or oat bran.

Spirulina is one of these foods. In Japan, thirty male employees with high cholesterol, mild hypertension, and hyperlipidemia showed lower serum cholesterol, triglyceride and LDL (undesirable fat) levels after eating spirulina for eight weeks. These men did not change their diet, except adding spirulina.

Group A consumed 4.2 grams (about 8 tablets) daily for eight weeks. Total serum cholesterol dropped a significant 4.5% within four weeks from 244 to 233. Group B consumed spirulina for four weeks, then stopped. Serum cholesterol decreased but then returned to the initial level. Researchers found triglyceride levels decreased slightly and LDL cholesterol decreased a significant 6.1% within four weeks. The reduction of serum cholesterol was even greater in those men with the highest cholesterol levels.

This study conducted by the Department of Internal Medicine of Tokai University concluded spirulina did lower serum cholesterol and was likely to have a favorable effect on alleviating heart disease since the arterioscelosis index improved. No adverse effects were noted. The study did not speculate on how it lowered cholesterol.[7]

Researchers in West Germany had previously discovered cholesterol reduction during a weight loss study with spirulina.[8] Japanese research showed lower cholesterol without weight loss, suggesting that cholesterol reduction was not related to weight loss. Spirulina was chosen because it previously lowered serum cholesterol in rats.[9,10]

A recent study with rats attempted to find the compound in spirulina that lowered serum cholesterol. Researchers discovered that the benefit may be through its effect on metabolism of lipoproteins. The oil soluble portion was found to suppress cholesterol levels in the serum and liver of rats.[11]

Natural beta carotene and cancer prevention

Cancer is the number two cause of premature death in Americans. Increasing cancer rates seem to be caused by environmental factors, especially diet. Scientists are examining foods and substances having protective factors. Beta carotene is one of the most well known natural anti-cancer substances. Over the past 20 years, cancer health authorities, National Cancer Institute and dozens of publicized studies have shown evidence that eating vegetables rich in beta carotene reduces the risk of all kinds of cancer.

Beta carotene is the main source of Vitamin A for humans. Our bodies convert beta carotene to Vitamin A as we need it. Although very high dosages of Vitamin A supplements are toxic, high amounts of beta carotene from foods and supplements are safe. Spirulina is the richest beta carotene food known, having over ten times more beta carotene than any other food, including carrots.

Beta carotene is one of the most effective substances for deactivating free radicals, which damage cells, leading to cancer. Free radicals are molecular fragments from environmental pollution, toxic chemicals, drugs, and physical and emotional stress. Beta carotene prevents free radicals from reacting, and decreases incidence of lung cancer, prevents chemically induced tumors in animals, prevents precancerous prechromosome damage and enhances immunological resistance.

Over 100 animal studies confirm Vitamin A and beta carotene inhibit the development of various cancers and tumors. Many human epidemiological studies correlated high Vitamin A intake with decreased cancer risks.[12] Beta carotene (and not the preformed Vitamin A from animal sources) correlated with lower cancer rates.[13]

Over 15 studies from 1975-1986 correlated lower incidence of lung cancer with beta carotene and Vitamin A. One study found the lower the serum level of beta carotene, the higher incidence of lung cancer.[14] Nine studies from 1974-1986 correlated lower digestive tract cancer (oral, stomach, colon, gastrointestinal) with beta carotene and Vitamin A. Two studies with women correlated lower breast and cervix cancer. A five year study in China completed in 1993 with 29,000 people revealed daily doses of beta carotene, vitamin E and selenium reduced the incidence of cancer deaths by 13%.[15]

In 1982, the famous monograph *Diet, Nutrition and Cancer* published by the US National Research Council reviewed this overwhelming literature. It concluded "the epidemiological evidence is sufficient to suggest that foods rich in carotenes or Vitamin A are associated with a reduced risk of cancer." The study recommended a diet including beta carotene rich vegetables to reduce cancer risks.[16]

A 1987 Israeli study demonstrated natural beta carotene is more effective than synthetic. Natural beta carotene is better assimilated by the body because it contains the 9-cis carotenoid isomer, lacking in synthetic carotene molecules. This means beta carotene in algae and vegetables greater antioxidant power than synthetic beta carotene.[17]

Controversy arose in 1995 when synthetic beta carotene was found ineffective preventing cancer in Finnish and U.S. smokers, and could even be harmful. Yet, these studies were flawed. Researchers chose only synthetic beta carotene lacking the cis isomer and gave high megadoses which may have caused nutrient imbalance. These studies reinforced the interest in natural carotenoids in whole foods.

Anti-cancer tumor effects

Because it is the richest natural beta carotene food, spirulina has been tested for anti-cancer effects. The Harvard University School of Dental Medicine reduced oral cancer cells with spirulina extracts. A beta carotene solution applied to cancerous tumors in mouths of hamsters reduced the number and size of tumors or caused them to disappear.[18] When a beta carotene extract was fed to 20 hamsters pre-treated to develop mouth cancer, none developed the disease. Tissue samples contained an immune stimulating substance believed to have destroyed cancer cells before they could multiply.[19]

In 1995, spirulina reversed oral cancer in pan tobacco chewers in Kerala, India. Complete regression of oral leukoplakia was found in 45% of those using one gram a day for one year, compared to only 7% with a placebo. Within one year of discontinuing spirulina, 45% of the lesions returned. This was the first human study of its chemo-preventive potential.[20]

Evidence linking natural beta carotene and cancer prevention is impressive. For those who do not eat 4-9 servings of fruits and vegetables daily, spirulina will add natural carotene insurance.

4.5. Summary of Clinical Research with Spirulina

Medical Subject	Patient	Country	Year
Anti-Aids viral effect	human cells	USA	1996[1]
Anti-viral effect	human cells	Japan	1996[2]
Anti-herpes viral effect	hamsters	Japan	1993[3]
Lowering cholesterol	humans	Japan	1988[7]
Lowering cholesterol	rats	India	1983[9]
Lowering cholesterol	rats	Japan	1984[10],90[11]
Reducing oral cancer tumors	hamsters	USA	1986[18],88[19]
Reducing oral cancer tumors	humans	India	1995[20]
Anti-liver cancer tumor	mice	Japan	1982[21]
Builds bone marrow	mice	China	1994[23]
Immune enhancement	rabbits	Russia	1979[24]
Immune enhancement	mice	Japan	1994[4]
Immune enhancement	mice	China	1991[28],94[25,26]
Immune enhancement	chickens	USA	1995[29],96[30]
Immune enhancement	cats	USA	1996[31]
Reducing kidney poisons	rats	Japan	1988[37]
from drugs and heavy metals	rats	Japan	1990[38]
Effect against diabetes	rats	Japan	1991[39]
Reduces blood pressure	rats	Japan	1990[40]
Building healthy lactobacillus	rats	Japan	1987[41]
Treating external wounds	humans	France	1967[43]
Treating external wounds	humans	Japan	1977[44]
Infection (antibiotic action)	microbial cells	PuertoRico	1970[45]
Infection (antibiotic action)	microbial cells	India	1978[46]
Recovering from malnutrition	humans	Mexico	1973[47]
Recovering from malnutrition	humans	Togo	1986[49]
Treating nutritional deficiencies	humans	India	1991[51],93[50]
Treating nutritional deficiencies	humans	Romania	1984[52]
Treating nutritional deficiencies	humans	China	1987[53],94[54]
High iron bioavailability	rats	USA	1986[55]
Correcting iron anemia	rats	Japan	1982[56]
Correcting iron anemia	humans	Japan	1978[57]
Treating infirmities with GLA	humans	Spain	1987[66]
Weight lowering effect	humans	Germany	1986[70]
Radiation protective effects	mice	China	1989[74]
Reducing radionucleides	humans	Belarus	1993[71,73]
Reducing radiation allergies	humans	Russia	1994[75]

Phycocyanin enhances the immune system

Part of the global effort to identify natural substances with an immune system boosting or anti-cancer effect focuses on blue-green algae. One unusual phytonutrient in blue-green algae is the natural blue protein pigment, phycocyanin.

In research in Japan, phycocyanin was taken orally by mice with liver cancer. The survival rate of the treatment group was significantly higher than the control group not given phycocyanin. After five weeks, 90% of the phycocyanin group survived, but only 25% of the control group. After eight weeks, 25% of the phycocyanin group still survived, yet none of the control group was alive. This suggests eating phycocyanin may increase survival of cancer stricken organisms.

In another study, after two weeks the white blood cells (lymphocyte activity) of a phycocyanin group were higher than the control group and higher than or equal to a normal group without cancer. This suggests phycocyanin raises lymphocyte activity.[21]

The lymph system's general function is to maintain healthy organs in the body, and protect against cancer, ulcers, bleeding piles and other diseases. These results suggest phycocyanin acts not by a limited attack on local cancer, but by strengthening the body's resistance through the lymph system. Phycocyanin may be active in preventing a host of degenerative organ diseases by increasing immunity.

A Japanese patent states a small dosage of phycocyanin daily maintains or accelerates normal control cell functions that prevents generation of malignancy such as cancer or inhibits its growth or recurrence.[22] The patent recommends a phycocyanin dosage in a range of 0.25 to 2.5 grams per day. Spirulina from Earthrise Farms, California and Siam Algae, Thailand have a high 15% phycocyanin content. This means the patent recommended adult dosage would be filled by consuming 1.7 to 17 grams per day, making about ten grams a day a useful level.

Chinese scientists documented phycocyanin stimulates hematopoesis (creation of blood), emulating the hormone erythropoetin (EPO). EPO is produced by healthy kidneys and regulates bone marrow stem cell production of red blood cells. They claim phycocyanin regulated white blood cell production, even when bone marrow stem cells are damaged by toxic chemicals or radiation.[23]

Polysaccharides enhance the immune system

In 1979, Russian scientists published initial research on the immune stimulating effects on rabbits from lipopolysaccharides in spirulina.[24] More recent studies in China and Japan have shown polysaccharide extracts increased macrophage function, antibody production and infection fighting T-cells.

In 1991-94 in China, polysaccharides and phycocyanin from spirulina increased immunity in mice by enhancing bone marrow reproduction, growth of thymus and spleen and biosynthesis of serum protein.[25,26,27,28] In 1993 in Japan, hamsters treated with a polysaccharide extract had better recovery rates when infected with a herpes virus.[3] In 1996, a water extract unique to spirulina, *Calcium Spirulan*, inhibited replication of HIV-1, Herpes Simplex and other viruses, yet was very safe for human cells.[2]

In the USA, a water soluble extract increased macrophage activity in chickens. In further 1993-96 studies, chickens fed a diet with less than 1% spirulina showed improved immune performance without any adverse side effects The whole immune system array of killer cells, helper cells and antibody production was supercharged.[29,30] Similar benefits were found for cats.[31]

Researchers are testing the theory that spirulina and its extracts act much like a broad spectrum vaccine against bacteria. Because it is a safe natural food, this research has created a sensation among animal scientists. They are scrambling to replace ineffective antibiotics with probiotics that strengthen the immune system and prevent disease. Based on this animal research, as little as 3 grams per day may be effective for humans.[32]

In 1996, U.S. scientists announced on-going research, documenting that a water extract of spirulina inhibits HIV-1 replication in human derived T-cells and in human blood mononuclear cells.[1] HIV-1 is the AIDS virus. Small amounts of the extract reduced viral replication, while higher concentration totally stopped its reproduction. The extract seemed to prevent the virus from penetrating the cell membrane, therefore the virus was unable to replicate. This spirulina extract was non-toxic to human cells. The scientists said this was only preliminary research, to be followed by animal and human studies.

Sulfolipid extracts from blue-green algae stop the HIV virus

The Natural Products Branch of the National Cancer Institute (NCI) is searching the world for natural plants and organisms that have biologically active anti-cancer agents. The famous periwinkle plant in the Madagascar rain forests is one example of a new cancer cure. Having scoured terrestrial organisms, scientists are now looking towards the sea.

NCI scientists have screened 18,000 extracts of marine organisms for activity against tumors, viruses and fungi and for immune system stimulation properties. Extracts of sea squirts, sea whip soft corals, and sea sponges offer potential new drugs.

In 1986, the NCI began studying thousands of types of blue-green algae for effects against the AIDS virus and 100 types of cancer. In 1989, the NCI announced that chemicals from blue-green algae were found to be "remarkably active" against the AIDS virus.[33] These are the naturally occurring sulfolipid portions of the glycolipids. Sulfolipids can prevent viruses from either attaching to or penetrating into cells, thus preventing viral infection.

NCI emphasized that a larger testing program including tests on humans with the AIDS virus would not begin until sulfolipids can be obtained in much larger quantities. These scientists further speculated that if sulfolipids proved effective, used in combination with drugs like AZT, they would be safer and more effective.

Scientists used extracts of the blue-green algae lyngbya, phormidium, oscillatoria (a member of the spirulina family) and anabaena. Spirulina is known to contain glycolipids and sulfolipids.[34] It contains 5-8% lipids, and of that, about 40% are glycolipids, and 2-5% are sulfolipids.[35] Analysis by Earthrise Farms revealed it has about 1% sulfolipids. Blue-green algae can be cultivated in ways to significantly increase the lipids, and presumably, the sulfolipids. This means it could be grown on a large scale for extraction of this valuable anti-cancer and anti-AIDS substance.

In 1996, NCI scientists announced another extract from the blue-green algae nostoc, *cyanovirin-n*, could be a broad spectrum virucidal agent against HIV. This unique antiviral protein was selected for further high-priority preclinical development.[36]

Reduces kidney poisons from mercury and drugs

Kidneys play an essential role in cleansing the body of toxins. Heavy metals and many drugs are known to be toxic to the kidneys. Scientists are interested in substances that can help cleanse the kidneys of toxic side effects from heavy metal poisoning or from high intake of medicines or pharmaceutical drugs.

In Japan, spirulina reduced kidney nephrotoxicity from mercury and three pharmaceutical drugs in laboratory rats.[37] Scientists measured two indicators of kidney toxicity-blood urea nitrogen (BUN) and serum creatinine. When the rats were fed a diet with 30% spirulina, BUN and serum creatinine levels decreased dramatically.

Similar effects were observed when rats were given common drug medications: para-aminophenol (painkiller), Gentamicin (antibiotic) and cis-dichloro-diaminoplatinum (anti-cancer drug). In all cases, the spirulina diet greatly decreased BUN and serum creatinine levels, and in two cases, brought serum creatinine down to original levels.

In a follow-up study, urinary excretion of two enzymes were measured as further indicators of renal function. The activities of both were significantly reduced in the group fed 30% spirulina. The effective compound responsible for the suppression of renal toxicity was the water soluble extract, phycocyanin.[38]

These studies suggest spirulina may have a beneficial effect for humans suffering from heavy metal poisoning. They also suggest kidney side effects from pharmaceutical drugs may decrease when it is eaten along with the administration of drugs. Side effects limit the dosage of many drugs, slowing the recovery period. With clinical use in hospitals or with outpatients, higher dosage of such drugs and shorter recovery times may be possible. In any event, study of the kidney cleansing effect offers an insight into the cleansing effects people have reported while fasting.

Effects against diabetes and hypertension

Spirulina may have a positive effect against diabetes. A water soluble fraction was found to be effective in lowering the serum glucose level at fasting while the water-insoluble fraction suppressed glucose level at glucose loading.[39] It may also reduce blood pressure. In a recent study with rats, it was found to reduce high blood pressure.[40]

Builds healthy lactobacillus

Healthy lactobacillus in the intestines provides humans with three major benefits: better digestion and absorption, protection from infection, and stimulation of the immune system. For these reasons, many people take lactobacillus supplements.

Research in Japan showed spirulina increased lactobacillus in rats by three times over a control group. Feeding rats a diet with 5% spirulina for 100 days revealed 1) the weight of the caecum increased 13%, 2) lactobacillus increased 327%, and 3) Vitamin B1 inside the caecum increased 43%.[41] Since spirulina did not supply this additional B1, it improved overall B1 absorption. The study suggests eating spirulina should increase lactobacillus in humans and may increase efficient absorption of B1 and other vitamins from the entire diet.

All of this has implications for Acquired Immune Deficiency Syndrome (AIDS). Some AIDS researchers believe the inability to absorb nutrients in the intestines (malabsorption or malnutrition) can cause serious immune deficiency. The absence of lactobacillus leads to thriving infections. In AIDS patients, nutrient malabsorption associated with 'opportunistic infections' can initiate full-blown AIDS.

One strategy for halting the progression of AIDS is based on nutrient supplementation (to correct malabsorption) and supplemental lactobacillus (to maintain proper intestinal flora and prevent infection).[42]

Wound healing and antibiotic effects

People have used spirulina in facial creams and body wraps, and there are reports of people taking it in baths to promote skin health. The Kanembu people in Chad have used the freshly harvested algae as a skin poultice for treating certain diseases.

In France, pharmaceutical compounds containing spirulina as an active ingredient accelerated wound healing. Several patients used whole spirulina, raw juice and extracts. Treatments consisted of creams, ointments, solutions and suspensions.[43] A study in Japan showed cosmetic packs containing spirulina and its enzymatic hydrolyzates promoted skin metabolism and reduced scars.[44]

Additional research showed extracts of spirulina inhibited the growth of bacteria, yeast and fungi.[45,46] The antibiotic substances in these extracts may have medical applications.

Benefits for malnourished children

As little a ten grams a day brings rapid recovery from malnutrition, especially for infants. Spirulina was given to undernourished children in Mexico,[47] and adults[48] with beneficial results. It was more than 10% of their diet and no adverse effects were noted.

In Togo, rapid recovery of malnourished infants was reported in a village clinic. Children given 10 to 15 grams per day mixed with millet, water and spices, recovered in several weeks.[49] In India, large scale studies with preschool children showed carotenes in spirulina helped children recover from symptoms of Vitamin A deficiency.[50,51]

In Romania, tablets were given to patients with nutritional deficiencies in a Bucharest municipal clinic. Patients had suffered weight loss in conjunction with chronic pancreatitis, rheumatoid arthritis, anemia, diabetes and other symptoms. The patients gained weight and their health improved.[52]

In China, spirulina was prescribed at Nanjing Children's Hospital as a 'baby nourishing formula' with baked barley sprouts. 27 of 30 children aged two to six recovered in a short period from bad appetite, night sweat, diarrhea and constipation. The researchers concluded this is a genuine health food for children.[53] In another study, children deficient in the essential mineral zinc, made more rapid recovery with high zinc spirulina than a standard zinc supplement.[54]

Iron bioavailability and correction of anemia

Iron is the most common mineral deficiency worldwide. Iron anemia is prevalent in women, children, older people, and especially women on weight loss diets. Iron is essential for healthy red blood cells and a strong immune system, but typical iron supplements are not well absorbed by the human body. Because spirulina is known to have a very high iron content, it was tested against a typical iron supplement. Spirulina fed rats absorbed 60% more iron than rats fed the iron supplement, suggesting there is a highly available form of iron in spirulina.[55] An earlier study showed it corrected anemia in rats.[56]

In Japan, eight young women had been limiting their meals to stay thin, and showed hypochronic anemia – lower than normal blood hemoglobin content. After four grams of spirulina after each meal, in 30 days, blood hemoglobin content increased 21% from 10.9 to 13.2, a satisfactory level, no longer considered anemic.[57]

GLA and prostaglandin stimulation

Foods high in saturated fats, typical of the American diet, may block the beneficial work of essential fatty acids in the human body, leading to many disease conditions.

Gamma linolenic acid (GLA), an essential fatty acid, is a precursor for the body's prostaglandins, master hormones that control many body functions. The prostaglandin PGE1 is involved in many tasks including regulation of blood pressure, cholesterol synthesis, inflammation and cell proliferation. PGE1 is usually formed from dietary linolenic acid, and the GLA progresses to PGE1.[58] Dietary saturated fats and alcohol and other factors may inhibit this process, resulting in GLA deficiency and suppressed PGE1 formation.[59]

Numerous studies have shown GLA deficiency may figure in degenerative diseases and other health problems. Clinical studies show dietary intake of GLA can help arthritis,[60] heart disease,[61] obesity[62] and zinc deficiency.[63] Alcoholism, manic-depression, aging symptoms and schizophrenia also have been ascribed partially to GLA deficiency.[64] A source of dietary GLA may help conditions of heart disease, premenstrual stress, obesity, arthritis and alcoholism.[65] In Spain, the GLA in spirulina and evening primrose oil is prescribed for treatment of various chronic health problems.[66]

The few known sources of GLA include two foods, human milk and spirulina, and oil of the evening primrose plant, black currant and borage seeds. Ten grams of spirulina has over 100 mg of GLA. This high amount of GLA is well documented.[67,68,69] It is about 5% essential fatty acids and 20% of this is GLA.

Weight loss research

In a study in Germany in 1986, researchers used 15 human volunteers to test an appetite reducing effect. Obese outpatients, who were already following a weight reduction diet, took spirulina tablets before each meal three times a day for four weeks. In this double blind crossover study against a placebo, about 6 tablets three times a day over four weeks showed a small but statistically significant reduction of body weight. There was also a significant drop in serum cholesterol levels.[70]

Reduces effects of radiation for the Children of Chernobyl

Years after the Chernobyl disaster, four million people in Ukraine and Belarus live in dangerously radioactive areas. The water, soil and food over an 11,000 square mile area is contaminated. Over 160,000 children are victims of radiation poisoning, with birth defects, leukemia, cancer, thyroid disease, anemia, loss of vision and appetite and depressed immune system, now called "Chernobyl AIDS."

Doctors reported spirulina's health benefits for child victims of Chernobyl radiation. Spirulina reduced urine radioactivity levels by 50% in only 20 days. This result was achieved by giving 5 grams a day to children at the Minsk, Belarus Institute of Radiation Medicine. The Institute program treated 100 children every 20 days.

An unpublished 1993 report confirmed 1990-91 research, concluding "spirulina decreases radiation dose load received from food contaminated with radionucleides, Cesium-137 and Strontium-90. It is favorable for normalizing the adaptive potential of children's bodies in conditions of long-lived low dose radiation."[71]

Based on testing in 1990, the Belarus Ministry of Health concluded spirulina promotes the evacuation of radionucleides from the human body. No side effects were registered. The Ministry considered this food was advisable for the treatment of people subject to radiation effects, and requested additional donations from the Earthrise Company of California and Dainippon Ink & Chemicals of Japan.[72]

4.6. Nurse and child radiation victim, at a medical clinic in Belarus.

Previous research in China in 1989 demonstrated a natural polysaccharide extract of spirulina had a protective effect against gamma radiation in mice.[74] Subsequent research showed phycocyanin and polysaccharides enhanced the reproduction of bone marrow and cellular immunity.[26]

In a 1991 study of 49 kindergarten children aged 3 to 7 years old in Beryozova, spirulina was given to 49 children for 45 days. Doctors found T-cell suppressors and beneficial hormones rose, and in 83% of the children, radioactivity of the urine decreased.[73]

A Russian patent was awarded in 1994 for the use of spirulina as a medical food to reduce allergic reactions from radiation sickness. The patent was based on a study of 270 children living in highly radioactive areas. They had chronic radiation sickness and elevated levels of Immunoglobulin (IgE), a marker for high allergy sensitivity. Thirty five were prescribed 20 tablets per day (about 5 grams) for 45 days. Consuming spirulina lowered the levels of IgE in the blood, which in turn, normalized allergic sensitivities in the body.[75]

4.7. Check out http://www.earthrise.com.

Keep up with the all the new scientific discoveries on the Internet

The pace of scientific research on spirulina's health benefits is accelerating, as more scientists unlock its secrets. Dozens of exciting new studies are being published every year. Fortunately, you can stay informed about these studies and their implications for your health on the World Wide Web. The Spirulina Health Library is one of the world's best information sources, with hundreds of scientific and popular references, updated every month. It has over 70 pages of published scientific abstracts and continues to expand.

Chapter 5

The variety of products around the world

In over 40 countries, people are already familiar with tablets, powder and capsules. Spirulina is a featured ingredient in pasta, cookies, snack bars and juice bars. It's used in personal care products like skin creams and shampoos. Look for it in innovative pet supplements for fish, aquatic animals, birds, cats and dogs. Natural food colors from spirulina have been widely used in Japan for years. Coming soon are therapeutic extracts.

- Market evolution in the USA.
- Natural health food products around the world.
- Specialty feed for fish, aquaculture and animals.
- Pet food for exotic birds, cats and dogs.
- Natural colors for foods and cosmetics.
- Fluorescent markers for medical tests.
- Enzymes for genetic research.
- New super spirulina products and extracts.

Market evolution in the USA

For centuries, traditional peoples have eaten seaweeds and aquatic vegetables. The first microalgae cultivated and sold in industrialized society was chlorella. Beginning in the 1970s, thousands of tons have been sold as health food supplements. Chlorella offers many health benefits, backed by decades of clinical research.

Japanese culture has always been receptive to sea products. Spirulina began selling in Japan as a health food supplement in the late 1970s. In the USA, in 1979, it was introduced through natural food stores by the Earthrise Company, and through a multi-level sales network by the Light Force Company.

Introduction, 1979-1981

Would Americans eat algae? The natural food market seemed a good place to ask. In the 1970s about 10% of all U.S. consumers were already eating some health or natural foods. They chose food based on nutritional information, not simply inherited cultural traditions. Concerned about the decline in food quality from over-processing, chemical additives, toxic pesticides and pollutants, they opted for natural foods over synthetic supplements.

Natural food customers quickly embraced this new algae, discovering many new ways to use it in their diet. Spirulina gathered a small, dedicated and steadily growing following, but remained a long way from mainstream recognition.

Diet boom, 1981-1983

In 1981, a national tabloid's front page headlines announced: *Doctor's Praise: A Safe Diet Pill – You'll Never Go Hungry.*[1] The article claimed spirulina was a safe and effective appetite suppressant. The very next day, thousands of people who had never been inside a health food store, lined up outside to try out the latest 'magic' diet pill. Spirulina became a new diet phenomenon. News was passed by word of mouth, by magazines and newspapers, by radio and television.

Hundreds of diet pill companies jumped on the bandwagon, rushing their own brand to market. Within a month, new brands appeared in health food stores, drug stores and supermarkets across America.

But there wasn't enough to go around. In 1981, the entire world supply came from Mexico and Thailand, and this was only about 500 tons a year. Most spirulina grown in Thailand was sold in Japan and most of the product grown in Mexico was already contracted to fish feed companies in Japan. Only several hundred tons per year of authentic spirulina actually entered the US market.

Popularity subsides, 1984-1986

Because of this supply shortage, many new brands sold by diet pill companies were adulterated, having less than 50% of the amount claimed on the label. Some products contained none at all. A lot was adulterated with alfalfa or cheap green fillers. The media hype led most people to believe it was nothing but a diet product. Hoping to magically lose weight, most people paid for green pills that were not even the real thing, and they experienced little health benefit. Spirulina hit bottom by 1986.

Coming back, 1987-1990

Natural food consumers who knew how to use it properly and enjoy its benefits remained faithful. As more people recognized the real health value, demand steadily rose again, along with the other green superfoods such as chlorella (green algae) and aphanizomenon (blue-green algae), barley grass and wheat grass.

Renewed growth, 1991- present

Green superfoods have been growing in popularity about 30-40% per year since 1991. New spirulina products appeared, formulated with natural herbs, phytonutrients, vitamins and minerals to raise energy levels, reduce pre-menstrual stress, improve athletic performance and endurance, develop a lighter appetite, offer anti-oxidant protection. Popular meal replacements with spirulina offered chocolate, fruit and vanilla flavors. Other companies invented snack bars, pasta, and fruit and vegetable juices.

The public has become more aware that nutraceutical foods have health benefits beyond simple nutrition. Recently people have learned about whole foods with phytonutrient compounds than can help prevent disease. Year by year, as more research on spirulina's health benefits are published, it becomes better known as a nutraceutical.

Spirulina products around the world

5.2. Spain.
5.3. Slovenia.
5.4. Taiwan.
5.5. United States.
5.6. Japan – chips.
5.7. Japan – crackers.

태양에너지·공기· 바다의 신비로 녹색의
혁명을 일으킨 지구 스피루리나 !!

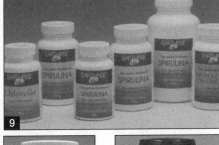

5.8. *Korea.*
5.9. *Germany.*
5.10. *Venezuela.*
5.11. *Japan.*
5.12. *Costa Rica.*
5.13. *Spain.*

85

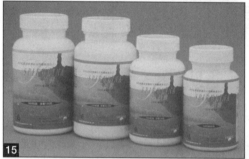

5.14. *Slovenia – dips.*
5.15. *China.*
5.16. *Germany – pasta.*
5.17. *New Zealand.*
5.18. *Canada.*
5.19. *Korea – soap.*

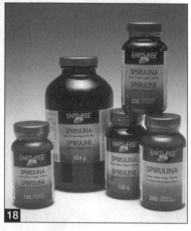

A specialty food
for fish, birds, animals and pets

Demand is surging for specialty aquaculture feeds that increase growth rates and disease resistance for farmed fish and prawns. Tropical fish, ornamental birds, animals and pets of all kinds consume much of the global spirulina production.

Color beauty for champion koi

The first modern use was to color the highly prized fancy koi carp in the Orient. These beautiful fish have distinctive bright red, yellow, orange, white and black markings and are often seen swimming in ornamental pools and fountains. They have been bred by hobbyists and fish culturists in Japan and China for centuries. If properly taken care of, koi can live longer than people, becoming family pets for several generations.

Champion fish will sell for tens of thousands of dollars each, so feeding them the right food is a good investment. Koi feeds include 5 to 20% spirulina for its rich carotene pigments that enhance the red and yellow patterns, while leaving a brilliant pure white. This clarity and color definition increases their value.

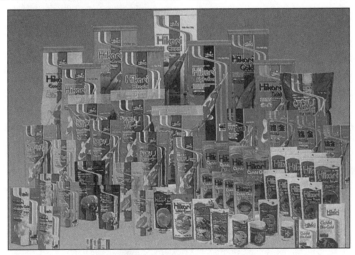

5.20. A popular brand of fish food for koi, ciclids and other colorfish contains spirulina as a key ingredient.

5.21. Spirulina – Benefits for aquaculture

Species	Benefits
Aquariums	Provides the base link of the food chain for a thriving tank and zooplankton community.
Ayu fish	Improves flavor and texture of meat, improves color, reduces disease.
Baby fish	Increased resistance to parasites, viral and bacterial diseases due to enhanced cellular immune system function. Stimulates feeding, better survival, improved appearance.
Brine shrimp	Replaces live algae for growth. Use as enrichment before feeding to fish or prawns. Improves color.
Fancy koi carp	Enhances skin quality, color, shine and disease resistance. Improves appearance.
Fish & marine larvae	Increases appetite, nutrition, growth and survival rates.
Milkfish	Improves growth and reproduction.
Prawns	Replaces live algae. Increases resistance to disease, stimulates growth, improves reproduction and enhances color.
Queen conch	Improves growth in nursery, reduces need for live algae.
Salmon	Enhances disease resistance, improves appetite, skin quality, color, reproduction.
Tilapia	Protects from parasite infection and bacterial disease. Enhances cellular immune system function, coloration and appetite.
Tropical fish	Enhances skin quality, color, shine and disease resistance. Improves appearance.
Yellowtail tuna	Reduces disease, increases growth and survival. Enhances coloration.

Courtesy of Ronald Henson. PO Box 459, Tollhouse CA 93667 USA.

Popular for aquaculture grown products

Fishing fleets are rapidly depleting ocean fish, threatening the ocean food chains. To offset the dwindling wild catch, the fish farming industry is growing at 10% per year. Adding spirulina to fish feeds helps solve the two biggest problems for growers. First, farmed fish are susceptible to infection and disease. Second, the flavor, texture and skin color are often inferior to the wild fish.

Japan is the largest market for aquaculture grown fish. Growers of ayu, a sweet fresh water fish, know spirulina improves skin color, growth rate and even the smell. Growers of yellowtail tuna, a popular sushi fish, know it increases survival and growth rates of the fish hatchery fry and improves the yellow side lines, skin color and growth rate. Prawn growers have discovered improved health. Prawns are fed spirulina just before harvesting to enhance their splendid colors for appeal in sushi bars.

Japanese fish farmers discovered five key benefits to using feeds with spirulina: 1) better growth rates, 2) improved quality and coloration, 3) better survival rates, 4) reduced medication requirements, 5) and reduced waste in the effluent.[3]

Growth rates increase and less feed is wasted because spirulina increases palatability of the entire feed. Fish respond to its flavor, and have less abdominal fat. Fish grow faster, taste better and resist disease. Environmental restrictions make it vital to reduce effluent pollutants. This makes it easier for fish farms to improve effluent quality without costly treatment systems. These are ways spirulina improves the cost/performance ratio of fish feeds.

5.22. Powder is a popular supplement for hatcheries.

Improves survival of fish fry and brine shrimp

Aqaculture fish are grown from tiny hatchery fry. This first stage is difficult and critical for success in aquaculture. Often, survival rates are very low. Spirulina added to the feed ration at 1 to 10% levels increases survival rates, allowing fish to reach market size sooner. It is the best food for tiny brine shrimp, sold in pet stores as a popular food for aquarium fish. Tiny zooplankton are another delicious food for larger fish, but growers have found them hard to cultivate. When fed spirulina, chances of successful cultivation improve greatly.

Health food for the aquarium

For many species of exotic tropical fish, algae are an essential part of the diet. Spirulina promises five benefits for healthy aquarium fish: 1) great profile of natural vitamins and minerals, 2) rich in muco-proteins for healthy skin, 3) phycocyanin for reduced obesity and better health, 4) essential fatty acids for proper organ development, 5) rich in natural coloring agents such as carotenoids. Feeding this algae will result in beautiful, healthy and longer lived tropical fish.[2]

Spirulina will not grow in the aquarium tank. Hobbyists can find floating flake food enriched with spirulina in many pet shops. Tablets dropped into an aquarium can create excitement. Some hobbyists use frozen food or make their own homemade fish foods. Public aquariums use spirulina to feed marine fish and invertebrates, and to raise daphnia and brine shrimp to feed to their fish.

Health, beauty and color for ornamental birds

Zoos around the world feed flamingos and ibis a diet rich in spirulina, and report major improvement in health and color. Algae increases feather color and shine, healthy beaks and skin, and promotes good bacteria in the digestive tract. Birds appear healthier, without synthetic vitamins, drugs or chemicals.[4] For hobbyists, pet birds can be more beautiful, healthier and live longer. Spirulina is very concentrated; use at 1 or 2% of total diet. Ailing birds may be fed more, not exceeding 5% of diet. Sprinkle powder over soft food or mix dry into baby hand feeding formula.

5.23. Enhances bird beauty and health.

Enhanced fertility and health for bird breeders

Canary, finch, parrot, lovebird and other breeders use spirulina to increase coloration, accelerate growth and sexual maturity and improve fertility rates. It is used by ostrich and turkey breeders to increase fertility and reproduction rates. It enhances desirable yellow skin coloration in chickens and increases the deep yellow color of egg yolks. Studies with chickens suggest that adding a small percentage in the diet stimulates macrophage production, improving immune performance and disease resistance without side effects.[5]

5.24. Becoming a popular superfood for birds, cats and dogs.

Healthy food for cats and dogs

For healthy skin and lustrous coat, spirulina is an excellent supplement for all cats and dogs, especially for nursing mothers and bottlefed kittens and puppies. Appetites of finicky cats have been known to perk up with a little sprinkled on their food. Owners report spirulina fed pets have a fresher breath odor.

Use a spoonful of powder in an empty salt shaker and sprinkle over soft moist food. Very concentrated. A little goes a long way. Use at about 1% of dry weight of daily food intake. Just a pinch each day for small cats and kittens; for large cats and nursing mothers use 1/4 teaspoon. For small dogs use 1/4 teaspoon each day; medium dogs use 1/2; large dogs use 1; giant dogs, 2 or more.

Use 1/4 or 1/2 tablet a day for small cats and kittens; large cats and nursing mothers use 1 tablet. For small dogs use 1/2 or 1 tablet; medium dogs use 2 tablets; large dogs 3-4 tablets; giant dogs, 4 or more.

Tonic for horses, cows and breeding bulls

Owners of highly valued racehorses use spirulina in their feed ration for faster times and recovery, but trainers tend to keep results secret. Dairy farmers use it to keep cows healthy by improving intestinal flora that is so important in ruminants. As a tonic for horses and cows, 1/2 ounce may be used twice a day for each 100 pounds of body weight by mixing in slightly dampened feed. Reports have circulated that it increases the sperm count of breeding bulls and fertility in females, raising the reproduction rates of valued species.

5.25. Spirulina—Benefits for pets and animals

Pets	Benefits
Bird	Improves feather quality, color, fertility, speeds development of immune system in babies for better resistance to infection.
Cat	Improves coat, healthy skin, healing, resistance to viral infections and cancer.
Dog	Improves coat, healthy skin, reduces dermatitis, improves disease resistance.
Horse	Reduces anemia, builds red and white blood cells, improves appetite, energy, resistance to infection, recovery from injury. Shiny coat.
Parrot	Improves feather quality, color, fertility, speeds development of immune system in babies for better resistance to infection.
Rabbit	Improves resistance to respiratory infection.
Reptile	Improves skin gloss, cleaner shedding, disease resistance.
Animals	Benefits
Dairy Cattle	Short-term therapeutic for mastitis, stimulates appetite.
Poultry	Reduces early poultry mortality.
Swine	Improves disease resistance in weaner pigs.

Courtesy of Ronald Henson. PO Box 459, Tollhouse CA 93667 USA.

Natural colors for foods and cosmetics

If you take a half teaspoon of powder and mix it in water, within an hour the particles will settle to the bottom, and a beautiful blue color will appear in the water. From another angle it may show a red or violet fluorescence. This water-soluble color is phycocyanin.

Spirulina contains 10-20% phycocyanin by weight and this makes it the best natural blue color for foods. Although you may not see blue food very often, blue, red and yellow are the three primary colors, and blue is used to make other appealing colors.

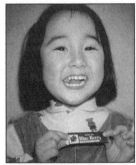

In Japan, food regulations mandate the use, whenever possible, of natural food colors without chemicals, solvents or preservatives. Dainippon Ink & Chemicals has developed a blue food color from spirulina called Lina-Blue®, used in chewing gum, ice sherberts, popsicles, candies, soft drinks, dairy products and wasabi, the green colored hot paste served in sushi bars.[6] Another form is made for natural cosmetics.

5.26. Yuki enjoys spirulina colored chewing gum (courtesy H. Shimamatsu). 5.27. Japanese chewing gum advertisement.

Spirulina is the best source of chlorophyll-a, and it is extracted for a green food color. Chlorella, has more chlorophyll-b, a grayer green color extracted and used in mouthwashes and deodorants.

In the U.S., the Food and Drug Administration regulates food colors and additives. The FDA requires a long and expensive approval process before natural colors can be used in foods and cosmetics, making it difficult for natural colors to replace chemical food dyes. Europeans are switching from chemical to natural food colors, and natural blue from spirulina is available there.

Fluorescent markers for medical tests

Purified algae extracts are used to track diseases in the human body. These phycobiliproteins are highly sensitive fluorescent dyes. They improve detection of cancer, screen donated blood for AIDS and monitor blood levels of drugs. They enable simultaneous tracking of several diseases in a single sample, allowing easier diagnoses.

Phycobiliproteins are natural pigments in red and blue-green algae, including spirulina. They have names like allophycocyanin, B-phycoerythrin, R-phycoerythrin, C-phycoerythrin and C-phycocyanin. When purified, they are fluorescent, stable, and water soluble.

Researchers watch these fluorescent markers under a microscope, and follow their path inside the human body. They can be easily linked to molecules such as monoclonal antibodies. When these antibodies attach to receptor sites on cells or tissues, the cells can be viewed by the glowing dye marker illuminated by a beam of light.[7]

Prices for purified phycobiliproteins range from $8,000 to $40,000 per gram. They are used in medical diagnostic kits and fluorescence-activated cell sorter (FACS) instruments in hospital laboratories, but are prized because they are 10-30 times more intense than conventional dyes and are an alternative to unpopular radioactive markers.[8]

Enzymes for genetic research

Live spirulina contains many enzymes. Among these, three restriction enzymes have been discovered. Restriction enzymes (endonucleases) work like scissors to cut the DNA of invading enemy microbes. Researchers in genetic engineering use restriction enzymes to cut DNA at precise locations.[9]

One unique restriction enzyme in spirulina called *Spl-1* is not found in any other microbe, bacteria, fungi or algae. Japanese scientists extract *Spl-1* from living spirulina, and sell it as a reagent for genetic research in laboratories and institutes. Other enzymes are extracted from the blue-green algae anabaena and nostoc.

One theory proposed to explain spirulina's long life is the role of restriction enzymes. By cutting the DNA of invading enemy microbes, spirulina is thought to protect itself from threatening bacteria or viruses. The three identified restriction enzymes may have been good weapons against ecological enemies for 3.5 billion years.[10]

New technology brings 'super spirulina'

Super spirulina has arrived. Leading commercial farms are beginning to grow it with enhanced nutritional attributes. Because microalgae reproduces so rapidly, changes introduced while growing can dramatically increase certain phytochemicals.

Higher iron, zinc and trace minerals

Algae absorbs minerals added to the pond water while growing. This colloidal mineral complex is more bioavailable than synthetic mineral supplements. Since 1984, Earthrise Farms has produced spirulina with high iron content. Ten grams provide up to 55% of the U.S. Daily Value (DV) for iron. Far East Microalgae of Taiwan produces a high iron, high selenium and high zinc product.

Super spirulina with extra zinc and beta carotene has been developed by Earthrise Farms. A ten gram serving has 600% D.V. for beta carotene and 70% D.V. for zinc. This mineral is essential on a daily basis for good health, powering antioxidant enzymes and a strong immune system, yet most people get only half the zinc they need in their regular diet. Beta carotene and zinc work synergistically as antioxidants to protect against free radical cellular damage from pollution, chemical toxins, radiation and stress.

5.28. Super spirulina with higher zinc and beta carotene.

Higher phycocyanin and beta carotene

By carefully selecting the species, adapting the growing conditions, and using low temperature drying, some farms produce a higher phycocyanin content. Heat sterilization can destroy this prized natural blue pigment. Spirulina from Earthrise Farms in California and Siam Algae in Thailand contains 15 to 20% phycocyanin, and is the most desired for color extraction. Both Earthrise and Cyanotech of Hawaii produce spirulina with higher beta carotene.

Higher GLA and sulfolipids

Scientists want to increase specific essential fatty acid content, especially gamma-linolenic acid. GLA is 1% of spirulina, and comprises 20% of its lipid content. By modifying the growing conditions, scientists hope to raise GLA content to 5%, and at least double the sulfolipid content. Commercial products may eventually follow from the NCI research identifying anti-cancer and anti-AIDS effects of sulfolipids. If sulfolipids were extracted, the remainder would be a concentrated food with the protein, minerals and vitamins largely intact.

Polysaccharide extracts for fish and animals

As the aquaculture industry expands, there is tremendous opportunity for probiotic specialty feeds. Several companies are already offering fish feeds with natural polysaccharides and beta glucans from yeast to enhance immune activity. These feeds would replace antibiotics used to control disease in farmed animals. There are two problems with antibiotics: 1) appearance of antibiotic resistant microorganisms, and 2) accumulation of antibiotics in fish.

With the recent discoveries of the benefits of polysaccharides in spirulina, researchers are studying the immune properties of polysaccharide extracts closely for potential therapeutic uses.

Phytochemical and pharmaceutical extracts

Scientists are searching for phytochemicals and pharmaceuticals in algae. Because numerous studies have shown therapeutic effects at low doses, extracts of the active ingredients are promising. Growers will learn how to enhance the concentration of these substances.

In some cases it may not be necessary to use the plant extract directly. One example is the extraction of phycocyanin for food coloring and for immunofluorescence studies. In this case the algae and its extract must pass rigorous screening for toxicity before approval is granted for use.

A leading expert in large scale cultivation, Dr. Amha Belay, has suggested that if pharmaceutical extracts of algae prove useful, spirulina has advantages over other algae: 1) The technology for mass cultivation and harvesting is available. 2) It has two decades of toxicity testing and safe human use for centuries. 3) It is already sold in many parts of the world as a safe food supplement. 4) Microbial and heavy metal standards have been established. 5) Its protein can be used after extraction of desired pharmaceuticals.[11]

Recently, several pharmaceutical drug companies have become interested in growing algae for extracts to be used in new pharmaceutical drugs. Cultivation biotechnology at existing commercial farms and new photobioreactor systems under development will produce these emerging products.

5.29. Liquid extract sold in Japan.

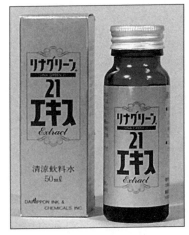

Chapter 6

How spirulina is ecologically grown

Blue-green algae growing in natural lakes may consist of several species. Harvesting can pose safety problems if one of them is toxic. However, by growing spirulina in designed ponds under controlled conditions, a pure culture can be maintained, which is not possible in natural lakes.

For these reasons, farms have developed rapidly over the last 20 years. Commercial systems have evolved in the United States, Thailand, Taiwan, Japan, Mexico, China, India and other countries.

- Guided tour of the world's largest spirulina farm.
- Ecological pond cultivation.
- Continuous harvest, drying, packaging.
- Rigorous quality control program.
- Comparison with chlorella, aphanizomenon, dunaliella.
- Worldwide spirulina farms.
- Growing in tubes, bioreactors and microfarms.
- Integrated production farms.

A guided tour of Earthrise Farms

Beginning in 1977, the first U.S. algae entrepreneurs began testing pilot ponds. They chose California's Imperial Valley because of its hot desert sunshine and remote location far from urban pollution. In 1981, a unique partnership between these California entrepreneurs and Japanese Corporate intrapreneurs founded Earthrise Farms. They shared a common vision of microalgae's coming impact on the global economy. Earthrise Farms began production in 1982.

World's largest spirulina farm

Merging U.S. and Japanese innovation, technology and resources, Earthrise became the world's largest spirulina farm, expanding to cover the entire 108 acre site. and today supplies over 40 countries with the world's best known spirulina. In 1996, the farm produced nearly 500 metric tons of dry powder.

6.3. Sunset over Earthrise Farms in the sunny California desert.

Cultivating only one pure algae

Keeping out weed algae.

Hundreds of aquatic organisms can bloom in nutrient rich water in warm sunshine, just as in a natural lake or swimming pool. Unlike a garden, weeding out unwanted algae is a difficult task since this algae is microscopic. Preventing weed algae from taking over is the key to growing a pure culture.

6.4. Pure cultures in the farm laboratory.

Ecological pond management.

Conventional farmers kill weeds and pests in their fields with pesticides and herbicides, leaving residues in the environment, on farm workers, and in your food. Earthrise Farms scientists keep out weed algae without toxic chemicals using a specially designed pond system and balancing the pond ecology. Producing spirulina under these controlled conditions does not allow growth of contaminant or weed algae as in lakes and waterways.

6.5. The primary production ponds have food grade liners.
Each pond is 5000 square meters- larger than a football field.

Ecological pond cultivation

Ecological farming produces vital, healthy and unpolluted foods. Although there are differences in farming land crops and algae, similar ecological practices are followed at Earthrise Farms. In this remote and sunny part of California, far from cities, highways and airports, the air is clean. Mineral rich Colorado river water, which supplies seven states of the U.S. Southwest, is pumped through canals, to large settling ponds, through filters into the growing ponds.

6.6. Clean fresh water and nutrients are added daily to feed the algae. No pesticides or herbicides are ever used.

Highest quality nutrients
Carbon Dioxide is bubbled into the water.

All plants need carbon to grow. Plant leaves take in carbon from the carbon dioxide (CO_2) in the atmosphere, but algae need carbon in the water. Algae grows so quickly that atmospheric CO_2 cannot penetrate the water fast enough to sustain growth, so carbon must be added. The same high quality CO_2 used in carbonated drinking waters is pumped directly into the ponds.

Pure mineral nutrients feed growing spirulina.

Adding extra manure or organic matter directly into the water can foul the shallow ponds and disturb algae growth. Instead, clean, pure sources of mineral nutrients are used. Premium quality minerals like nitrogen, potassium, iron and essential trace elements nourish a consistent high quality spirulina.

6.7. Long paddlewheels mix the pond water for optimum growth.

A continuous fresh harvest

During the growing season, April through October, ponds are harvested every day. In the peak summer sun, harvesting occurs 24 hours a day, around the clock, to keep up with the explosive growth rate.

6.8. Pumps send algae rich water to the sealed harvest building.

A 15 minute journey from ponds to dry powder.

Spirulina takes a quick trip through the stainless steel harvest and drying system, never touched by human hands. The first screen filters out pond debris. The next screens harvest the microscopic algae. The nutrient rich water is recycled back to the ponds. The final filter thickens spirulina from green yogurt to green dough. It is still 80% water inside the cells and needs to be dehydrated immediately.

Quick drying preserves nutrients

Spirulina droplets are sprayed into the drying chamber to flash evaporate the water. Dry powder is exposed to heat for several seconds as it falls to the bottom. Then it is vacuumed into a collection hopper in the packaging room. This quick process preserves heat sensitive nutrients, pigments and enzymes.

6.9. Spray drying tower.

Sealed in special oxygen barrier containers

No preservatives, stabilizers or additives are used in drying, and it is never irradiated. This powder has a high content of heat-sensitive phycocyanin, attesting to the drying quality. Fresh dried spirulina can be stored for five years or more in special oxygen barrier containers, retaining nearly full beta carotene potency.

6.10. These containers are shipped from the Earthrise Farms warehouse all over the world.

Tableted and bottled finished products

Spirulina powder is directly compressed into tablets and sealed into both glass and plastic bottles for Earthrise® products at a special facility at Earthrise Farms.

6.11. Automated bottling line.

Pesticide free

Earthrise Farms has never used pesticides or herbicides. Because more and more people are concerned about pesticide residues in their foods, spirulina has been additionally tested for residual environmental residues since 1993. Independent lab tests have not detected any of over 66 possible contaminant residues, so product labels now state "pesticide free." Earthrise Farms felt this extra level of assurance was important to quality conscious natural food customers.

Certified organic

In 1996, Earthrise Farms began growing organic product, and will expand output based on market demand. A California state registered agency certified that it meets guidelines set by the U.S. Organic Foods Protection Act and the California Organic Foods Act of 1990. Organic foods must be produced with an approved farm plan which demonstrates agricultural sustainability through the use of acceptable nutrients, materials and crop management practices.

6.12. Juan Chavez, V.P., and certified organic growing pond, 1996.

Super spirulina and new research

Super spirulina has enhanced levels of beneficial nutrients. In 1994, Earthrise Farms launched spirulina containing 30 times more zinc, an essential mineral for our immune system. Organically bound zinc is more bioavailable than conventional zinc supplements. Farm scientists have been cooperating with researchers from outside institutions on scientific methodology, to document therapeutic properties and search for bioactive phytonutrients.

Rigorous quality control program

Earthrise Farms technicians collect samples from the pond water and the final dry product for dozens of different analytical tests. Only after each lot has passed all tests is it certified ready to ship, accompanied by a laboratory report. Additionally, independent laboratories in the USA and Japan periodically confirm protein, pigments, amino acids, vitamins, minerals, microbiology and heavy metals.

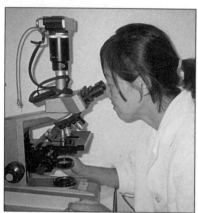 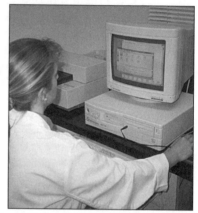

6.13-14. Technicians conduct 40 QC tests each day on living ponds and dry product.

Because of the concern about toxic blue-algae growing in lakes, Earthrise Farms developed a program to assure these toxic algae are not present in spirulina ponds. First, daily microscopic examinations of the living culture. Second, farm scientists developed in cooperation with university researchers, immunoassay and enzyme inhibition bioassay methods to detect nanogram levels of toxins. Very few laboratories in the world are capable of such analysis.

Meets all international food safety and quality guidelines

Earthrise Farms is subject to inspections by the U.S. Food and Drug Administration and the state of California. Its success in exceeding all food safety and quality standards is attributed to the remote farm location, clean air and water, premium quality nutrients, and the total quality management approach to quality control.

Comparing spirulina to other microalgae

Three other microalgae are sold as natural food supplements: chlorella (green algae), aphanizomenon flos-aquae (blue-green algae), and dunaliella (red algae).

Chlorella – green microalgae

In the 1960s, early research focused on chlorella. This green microalgae is a small spherical cell with a cell nucleus, and appeared a billion years after blue-green algae. This was the first commercial success for microalgae, cultivated by specially designed high tech commercial farms. Thousands of tons have been sold each year for the past 20 years. Farms in Taiwan, Southern Japan and Indonesia produce almost the entire world supply.

Three drawbacks limit its potential as a new food resource for the developing world. *First,* chlorella culture is easily contaminated by undesirable weed algae. Unlike spirulina which flourishes in highly alkaline water unfriendly to other algae, chlorella grows in normal water conditions where many algae grow. Chlorella is grown in individual batches, started in a sterile test tube, moved to indoor tanks and then outside to larger ponds. When it achieves maximum density, the entire batch is harvested. Controls required for batch cultivation are more difficult for less sophisticated developing world farms.

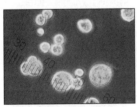

Second, unlike spirulina, tiny chlorella cells cannot be harvested by a screen. Expensive centrifuges are required to separate the cells from pond water. Spirulina can be harvested with a simple cloth or a fine screen.

6.15. Chlorella cells in the microscope.

Third, chlorella's hard cellulose cell wall protects its nucleus but resists digestion by the human body, and nutrients cannot be fully absorbed. Commercial farms crack open this hard cell wall in the drying process, or mechanically crush it. Cell breaking procedures would be costly for low technology farms.

When researchers rediscovered spirulina in the 1960s, they praised its nutritional value and ease of cultivation. Even though chlorella was cultivated first, many algae scientists soon afterwards forecast spirulina would become a food of the future.

"Wild" blue-green algae

Many people have asked about the differences between spirulina and wild blue-green algae harvested from lakes. Spirulina is one particular kind of blue-green algae with a centuries long history of safe human consumption. It is well known to be safe and nutritious. Hundreds of published scientific studies over the past thirty years have documented no toxicity.

Blue-green algae is also called cyanobacteria. Wild species grow in lakes and waterways, consuming whatever nutrients are in the water. Some species of cyanobacteria are toxic just like mushrooms and some land plants. Most kinds of microcystis, anabaena and aphanizomenon are toxic. Wild blue-green algae has not been subject to spirulina's long safety testing.

6.16. Aphanizomenon flos-aquae cells in the microscope.

Companies that harvest wild species of blue-green algae from natural lakes cannot have the same degree of control as growing spirulina at Earthrise Farms. Harvesting wild algae presents a far greater risk of contamination by cyanobacterial toxins.

Spirulina can be cultivated in a monoculture, uncontaminated by other cyanobacteria. Farms can be specially designed and operated to produce spirulina under controlled conditions that do not allow the growth of other contaminant cyanobacteria as in lakes and waterways. Finally, 40 quality control tests assure spirulina meets all international food safety and quality standards. It is already known to be safe and nutritious, and the technology and quality control at an advanced farm assures product purity and safety.

Dunaliella for beta carotene

Dunaliella is cultivated for its high content of beta carotene. Farms are located in Israel, California, Hawaii and Australia. Dunaliella likes hot climates and needs water even saltier than the ocean, making places like the Dead Sea in Israel a good location. This microalgae is too salty to be eaten directly as a whole food, but its beta carotene, over 10 times more concentrated than spirulina, is extracted as an oil or powder and sold as a natural food supplement and a color for aquaculture feeds.

Worldwide spirulina farms

Two kinds of outdoor farms operate today. The first harvests from an alkaline lake where spirulina is growing. The second are advanced outdoor pond cultivation systems. Under development are enclosed systems with transparent tubes, biocoils or photo-bioreactors.

Lake harvest farms

Lake harvest systems offer better quality control than harvesting wild algae, but have similar risks. They enjoy the advantages of inexpensive nutrient sources in the alkaline lakes and lower labor costs than developed countries, but quality may be inconsistent.

Mexico: In the 1970s, a Mexican company realized the algae in Lake Texcoco clogging the extraction of soda brines from the lake was spirulina. The world's first large plant was built here. Spirulina Mexicana has a larger potential capacity than any other farm.

6.17. Sosa Texcoco / Spirulina Mexicana near Mexico City.

In 1979, Mexican spirulina was first exported to the U.S. for use in health food products, but beginning in 1982, it was blocked by U.S. authorities due to quality problems. In following years, steps were taken to improve product quality, using heat sterilization to destroy bacteria levels from the lake. Lake Texcoco is located next to Mexico City, one of the most polluted cities. Higher levels of heavy metals in this product are a concern. Much of it is sold as animal and aquaculture feed. Although potentially the largest spirulina farm, it has not been operating for several years.

109

Myanmar: In 1988, commercial harvest began in Burma, now called Myanmar. Several alkaline volcanic lakes have natural blooms of spirulina. By 1993, 30 tons per year was being harvested, sun dried, tableted and sold on the local market (see Chapter 8).

Chad: The alkaline lakes around Lake Chad in Africa offer an ideal location. Various ventures been trying to produce algae near Lake Chad, but no commercial success has been reported.

Advanced pond cultivation systems

Most commercial farms are designed and built from the ground up. These farms have shallow raceway ponds circulated by paddle-wheels. They operate in a similar fashion to Earthrise Farms. Ponds vary in size up to 5000 square meters (about 1.25 acres), and water depth is usually 15 to 25 centimeters. These farms require more capital investment per area than lake systems, and operate at higher efficiency and quality control for a consistently high quality product.

6.18. Spirulina ponds of Siam Algae Company in Thailand.

Thailand: Dainippon Ink & Chemicals (DIC) started Siam Algae in 1978 near Bangkok, Thailand. With a tropical climate and a year-round growing season, Siam Algae has high productivity and grows 150 tons per year. Most is sold in Japan for health food products.

Hawaii, USA: Cyanotech opened a farm in 1985 on the Kona coast on the Big Island of Hawaii, producing both spirulina and dunaliella on a year round basis. Over the years this farm has expanded, and produces over 400 tons per year and sells all over the globe.

Taiwan: In the late 1970s, several chlorella farms were modified for spirulina. Five Taiwan farms have spirulina capability and can produce several hundred tons per year. Depending on market price, some shift to growing chlorella when its price is higher.

6.19. Yaeyama farm in Southern Japan grows chlorella in circular ponds.

Israel: The Desert Research Institute has researched spirulina for over 15 years. Several attempts at large scale production were started but did not achieve success.

China: Production began in 1987 in several research projects. Eight production farms were operating in South China and Hainan Island by 1993. By 1996, over 20 farms produced several hundred tons for humans and animals. China has become a major producer.

India: Spirulina research began in late 1970s in India, from backyard family scale to production farms. Two large commercial farms have an estimated capacity of 300 tons by 1996.

Vietnam: Experimental research began in the late 1970s, and limited production began in the 1980s. Since 1987, production has been about 5 tons per year, sold locally as health food and special feeds.

Chile: In 1991, Solarium started production in the Atacama desert. About 3 tons per year is produced and solar dried. Most is sold in the local market, but part is exported to other countries in South America.

Other farms: There are reports of production in Bangladesh, Cuba, Martinique, Peru, Brazil, Spain, Australia and other countries. Spirulina farms are blossoming around the world.

Tubes, bioreactors and microfarms

Spirulina needs hot temperatures to grow well. Most temperate climates are too cold for outdoor pond cultivation year round. This limits the locations where it can be grown economically. It may not even be possible to maintain a pure culture of other microalgae outdoors. To grow other algae and to increase growth rates, scientists have investigated new growing systems, called bioreactors.

Tube, coil and vertical plate systems

Plexiglass tubes and coils act as solar collectors, increasing temperature and extending the growing season. Algae is pumped continuously through rows of connected flexible transparent tubes or coils. Much greater density can be maintained than in open ponds.

6.20. A continuous loop photobioreactor for growing algae outdoors. (Courtesy of Tomaselli [3]).

Advantages are increased productivity, less water loss from evaporation, screening out contaminant algae, and greater control over the culture. On the downside, algae may stick to the inside of the tubes and block sunlight, and tubes may get too hot. Excessive oxygen produced by the algae while growing can reduce growth. A vertical plate system has been designed that has a flexible orientation to the sun, and allows oxygen to be released at the bottom.[2]

These systems are very expensive to build and have not yet been economically scaled up for commercial production to compete with open ponds, or for use in developing countries.

6.21. A flat plate reactor for growing spirulina outside. (Courtesy Tredici [2]).

Photo-bioreactors

These controlled indoor systems use transparent tanks and artificial lights. Bioreactors grow algae much like bacteria and yeast. Tubes and bioreactors are more expensive to build per unit area than outdoor ponds, but investors hope high-value pharmaceutical products will justify the costs. Sunlight and nutrients are the primary energy sources, but scientists are even looking into harnessing lasers. Laboratory scale systems use ultrasound and electromagnetic patterns to influence the growth of algae and increase certain attributes. These technologies are moving toward the closed ecological life support systems being tested for use in space stations.

Community microfarms

Ecologically designed solar powered communities of the future may incorporate an algae microfarm. Computers can handle the basic functions of cultivation inside the controlled greenhouse or bioreactor. On a tiny land area, a planned community could meet a significant portion of its protein and vitamin requirements from algae, freeing cropland for community recreation or reforestation.

Integrated production farms

Most advanced farms designed to produce high quality spirulina have necessarily higher production costs. To lower costs, future farms need to integrate sources of nutrients and energy, and produce a variety of end products, from valuable extracts to inexpensive protein.

The first company growing spirulina, the French Petroleum Institute, originally began cultivation next to an oil refinery using carbon dioxide gas byproduct to reduce production costs. This venture did not work for a variety of reasons. Nevertheless, the idea of using surplus or recycled nutrients is still very much alive.

Future farms may be sited on alkaline lakes in Africa where algae grows on free carbon nutrients. Other farms may locate near oil refineries or industrial centers using surplus industrial nutrients. Hot water from energy plants, or hot geothermal water, may provide heat to grow algae year-round in cooler climates. Using lower cost nutrients will lower production costs and will be attractive in the developing world or nutrient poor regions.

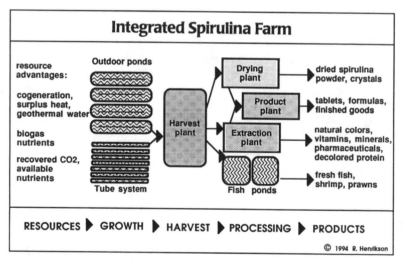

Carbon dioxide (CO_2) from biogas digesters fueled by plant, animal or human wastes can be recycled to grow spirulina. On the village level, this was achieved by the integrated system designed by Dr. Ripley Fox, described in Chapter 8.

Some farms will build integrated aquaculture systems. Asians have used algae to promote fish culture for centuries. Spirulina stimulates appetite, growth rate and disease resistance. Fresh wet algae can be added directly to fish ponds or to a dry feed ration. Even pond water rich with algae and saturated with oxygen can be pumped directly into fish ponds, delivering both food and oxygen.

Integrated farms will produce a variety of spirulina powder, tablets and bottled product. Some companies may specialize in pharmaceutical compounds, enzymes or medicines. Biochemical plants will make concentrated vitamin, fatty acid and pigment extracts. These valuable extractions would leave a 65% protein byproduct, much less expensive than regular spirulina powder.

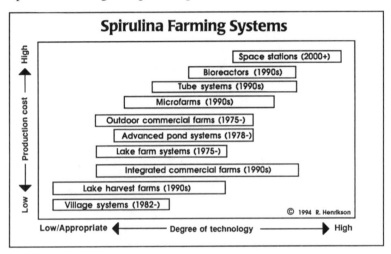

Blue-green algae is fairly simple to genetically alter. Some facilities may use genetic engineering to modify desired chemical compounds, induce the algae to grow faster and better in cold climates, or even fix nitrogen. Although this research holds promise, it is also cause for concern. Scientists cannot foresee all consequences and implications of modifying DNA in organisms.

If large commercial farms can use inexpensive resources and energy to grow spirulina, and if they can produce valuable extracts, the byproduct may become a much less costly food, becoming price competitive with other vegetable protein concentrates.

Chapter 7

Spirulina's resource advantages and world food politics

Conventional food production hides environmental costs. The cost accounting system doesn't take nature into account. It relentlessly destroys natural resources. You pay for these externalized costs, but not at the cash register.

Spirulina has no hidden environmental costs and offers more nutrition per acre than any other food. It conserves land and soil and uses water and energy more efficiently per kilo of protein than other foods. As global algae production expands using non-fertile land and brackish water, more cropland can be returned to forest. As more people eat lower on the food chain, we can halt pressure to destroy wilderness for cropland, and help regreen our planet.

- The hidden costs of food production.
- Spirulina resource advantages: Land, water and energy comparisons.
- Three problems facing global food production.
- Advantages of algae for food production.

The 'hidden costs' of food production

To gain perspective on the cost of spirulina relative to other foods, let's look at assumptions about the price of our food. Most people assume the store price reflects the real cost of producing food. Nothing could be farther from reality.

Agribusiness farming practices have externalized many production costs, and relentlessly destroy natural resources. You still pay for these costs, but not at the checkout counter. If you calculate these hidden costs of food you pay indirectly, and add these to the cash price, food prices would be much higher.

What are these hidden costs and how did they arise?

1. Medical costs from poisoned, unhealthy food.

Pesticides, fungicides, animal antibiotics, preservatives, chemical food additives, and fatty, salty foods all create long term health risks. The over processed foods promoted in advertising campaigns represent a very unhealthy diet. When the U.S. Surgeon General links two-thirds of all deaths to diet, this translates into higher medical bills, higher medical insurance, and higher taxes to support government health programs. Medical health care costs are the fastest growing sector of the U.S. economy.

2. Farm subsidies.

Taxes used for government farm subsidies support agribusiness and encourage wasteful consumption of water and soil. In some irrigation areas, the value of the crops grown with federal water is less than the cost of the water to grow these crops. At one water project, the full cost of water delivered was estimated to be $54 per acre foot, even though farmers were charged only $0.07 per acre foot.[1] In 1989, Farmers Home Administration bad loans were estimated at over $20 billion. These hidden subsidies and bad loan policies encourage waste and make food appear to be cheaper to produce and cheaper to buy.

3. Toxic cleanup costs.

Pesticides, herbicides and chemical fertilizers pollute our water and land. How much will these poisons hurt us, how long will they last, and what are the cleanup costs? We will pay much more tomorrow for cheap food today.

4. High global military costs.

Food choices made in the U.S. set the style for consumption across the entire world. To emulate the American meat-centered diet, developing world elites alter land use patterns in those countries. American green revolution agriculture and grain-fed beef farming methods tend to favor large, wealthy farmers. Because developing countries need cash to pay interest on their debt to our banks, they produce food to export, taking land away from local food production and displacing local farmers. This pattern creates food insecurity and chronically hungry people. One response to this human misery has been global militarization to maintain security in the face of exploitation.

5. Government debt and interest costs.

The hidden costs of food production in increased health services, military spending and farm subsidies have been gobbling up government revenues. The resulting budget deficit and interest costs pulls money away from investment in productive assets, blocking meaningful toxic cleanup and environmental restoration.

Cheap food is another aspect of our illusionary prosperity, pumped by a galloping national debt. When the U.S. devalues its currency and sells Treasury Bills to fund the national debt, it is selling and consuming real assets much too cheaply today in exchange for debt which must be paid tomorrow.

6. Environment and resource destruction.

Agribusiness treats fertile soil and precious fresh water like factory assets to be depreciated. The natural wealth is extracted, and not replaced. It accounts for rapid soil demineralization, salinization and erosion, the shocking drawdown of water aquifers, and the astounding loss of forests all over the world.

For example, does a fast food quarter pound hamburger cost only $2.49?

One quarter pound burger may come from U.S. grain fed beef. It takes 16 times as much corn to get protein from beef than from corn directly, and each pound of corn produced causes 2 pounds of topsoil erosion. An inch of topsoil takes 200 to 1000 years to form. Each 1/4 pound hamburger costs 8 pounds of irreplaceable American topsoil.[2] Soil and water alone may exceed the $2.49 price tag!

The burger that ate a rain forest

In Central America the cycle of destruction begins with the chainsaw, and ends a few hundred miles farther north in the | by about five cents a pound, and helped to reduce inflation.

The other side of the economic balance, | **It takes 55sq ft of rain forest to raise enough beef to make a single American hamburger**

7.2. "The burger that ate a rain forest" –London Times, Feb 26. 1989.

Or, burgers may come from imported beef. The U.S. imports 90% of all Central American beef exports for burgers, 138 million pounds of beef each year. Each burger really takes 55 square feet of tropical forest permanently cleared for grazing land. The burned vegetation emits 500 pounds of carbon dioxide into the atmosphere, aggravating the greenhouse effect.[3] That's not $2.49 either!

According to the Rainforest Action Network,[4] over half of the 5 billion acres of rainforest is gone. If present trends continue, it will be all gone in less than 40 years. In the Amazon, 6,000 rainforest clearing fires were burning out of control in 1988. Expanding cattle ranches have caused 72% of rainforest destruction in Brazil. Deforestation releases about a billion tons of carbon into the atmosphere each year. This represents one sixth of the total carbon release by human activity, adding to the greenhouse effect.

How much does this "burger that ate a rainforest" really cost? A fast food burger could cost $100, depending on how one values the Earth's resources. If everyone had the American appetite for beef, the entire planet would have to become one giant beef farm. A further discussion on these subjects is found in two books worth reading: *Diet For a Small Planet*[5] by Frances Moore Lappé, and *Diet for a New America*[6] by John Robbins.

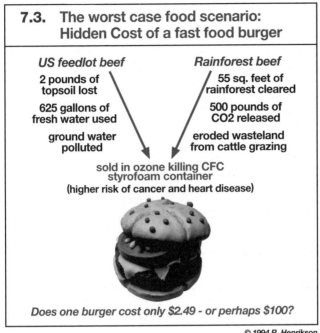

**7.3.   The worst case food scenario:
Hidden Cost of a fast food burger**

US feedlot beef

2 pounds of
topsoil lost

625 gallons of
fresh water used

ground water
polluted

Rainforest beef

55 sq. feet of
rainforest cleared

500 pounds of
CO2 released

eroded wasteland
from cattle grazing

sold in ozone killing CFC
styrofoam container
(higher risk of cancer and heart disease)

Does one burger cost only $2.49 - or perhaps $100?

© 1994 R. Henrikson

7. The accounting system ignores resource costs. It doesn't take nature into account.

We do not know what a fast food burger really costs because the economic accounting system simply ignores natural resource depletion and the concept of sustainable development. Gross National Product (GNP) figures and company balance sheets show man-made capital depreciation, but amazingly, not the consumption of precious soil, water, trees, minerals, fisheries or wildlife.

Accounting methods evolved many years ago when natural resources were considered free and unlimited. We need to begin *"taking nature into account,"* asserts the World Wildlife Fund. Only when the world economy shifts to natural resource accounting, such as the system developed by the World Resources Institute in *Wasting Assets, Natural Resources in the National Income Accounts,*[7] will we be able to measure the true cost of our food and all other products.

7.4. The Hidden Costs of Food

Hidden Costs	Spirulina[a] algaculture	Organic farming	Agribusiness crops	meat/ dairy
Poisons in foods (from production and processing)				
Pesticides	NO	NO	YES	YES
Herbicides	NO	NO	YES	YES
Preservatives	NO	NO	YES	YES
Additives	NO	NO	YES	YES
Animal antibiotics	NO	NO	NO	YES
Carcinogens	NO	NO	YES	YES
Higher medical costs (cancer, heart and organ diseases, degenerative disease)				
Your medical bills	NO	NO	YES	YES
Medical insurance	NO	NO	YES	YES
Gov't health payments (higher taxes)	NO	NO	YES	YES
Government farm subsidies				
(higher taxes)	NO	NO	YES	YES
Ecological destruction (from poor resource management)				
Loss of topsoil	NO	NO	YES	YES
Loss of fresh water	NO	NO	YES	YES
Loss of forests	NO	NO	YES	YES
Greenhouse effect	NO	NO	YES	YES
Loss of plant and animal species	NO	NO	YES	YES
Toxic cleanup costs coming due				
Soil pollution	NO	NO	YES	YES
Water pollution	NO	NO	YES	YES
Cleanup costs	NO	NO	YES	YES
Global militarization costs				
(higher taxes)	NO	NO	YES	YES
Higher debt burden				
(taxes and interest)	NO	NO	YES	YES
Exploitation				
Earth resources	NO	NO	YES	YES
Animal torture	NO	NO	NO	YES
Human suffering	NO	NO	YES	YES
Overall: ***Sustainable Path***	YES	YES	NO	NO

a. source: Earthrise Farms.

Spirulina prices reflect true costs

Like growing organic food, growing spirulina does not hide costs. Eating spirulina and organic foods will improve your health and lower your medical bills, compared to a diet rich in meat and conventionally grown foods. There are no big government subsidies for spirulina. Ecological cultivation does not cause pollution, soil erosion, water contamination or forest destruction.

Production costs range from $10 to $20 per kilo for commercial farms, depending on size and location. Farms with resource advantages like those in alkaline lakes may have lower production costs, ranging from $5 to $15 per kilo. Farms with year-round tropical growing seasons, energy and nutrient advantages, and extraction facilities for high-value products, may be able to produce a protein byproduct for a few dollars per kilo. This will become more price competitive with conventional proteins when the hidden costs of food production are taken into account. Whether or not the hidden costs are added in, spirulina production has resource advantages over conventional foods.

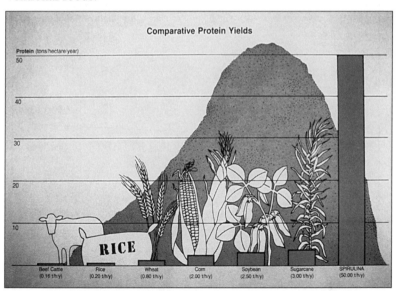

7.5. Comparative protein yields. (The Futurist, Feb. 1985)

Spirulina resource advantages
An environmentally sound green food revolution.

Spirulina cultivation at Earthrise Farms does not cause pollution, soil erosion, water contamination or forest destruction. Spirulina is grown without toxic pesticides and herbicides. These assurances are important to the conscious natural food customer.

Land and soil are conserved.

Spirulina is 60% protein and can be cultivated on marginal, unusable and non-fertile land. Its rapid growth means spirulina protein needs 20 times less land than soybeans, 40 times less than corn, and 200 times less than beef production. Spirulina offers more nutrition per acre than any other food, but doesn't even need fertile soil. Higher food production can be achieved while returning cropland to forest.

Land Area Needed to Produce One Kilogram of Protein

	Sq. Meters	Quality
Spirulina[a] 65% protein	0.6	non-fertile
Soybeans[b] 34% protein	16	fertile
Corn[b] 9% protein	22	fertile
Grain-fed Feedlot Beef[b] 20% protein	190	fertile

[a] Y. Ota, Earthrise Farms, California 1995
[b] Leesley, et al. "A low energy method of manufacturing high-grade protein using spirulina," University of Texas, 1980. Pimentel, 1975, USDA

One kilo of corn protein causes 22 kilos of topsoil loss. One kilo of protein from corn-fed beef is even more destructive, causing 145 kilos of topsoil loss from the cattle eating all that corn. Spirulina cultivation causes no topsoil erosion.

More efficient water use.

Even though spirulina grows in water, it uses far less water per kilo of protein than other common foods. At Earthrise Farms, water is recycled back to the ponds after harvesting. Because the main production ponds are sealed with food grade plastic liners, very little water seeps through the ground compared to conventional crops. The only significant loss is by evaporation.

Spirulina protein uses 1/3 the water as soy, 1/5 as corn, and only 1/50 the water needed for beef protein. At Earthrise Farms, only 8 gallons of water are used to produce a single 10 gram (20 tablet) serving of spirulina. This is much less than for servings of other foods grown in the U.S.: 15 gallons for a serving of bread, 65 for milk, 136 for eggs, 408 for chicken and 1303 gallons for a hamburger.[8]

Water Needed to Produce One Kilogram of Protein

		Liters	Quality
Spirulina[a] 65% protein	⬤	2100	brackish
Soybeans[b] 34% protein	⬤⬤⬤⬤	9000	fresh
Corn[b] 9% protein	⬤⬤⬤⬤⬤	12500	fresh
Grain-fed Feedlot Beef[b] 20% protein	⬤⬤⬤⬤⬤⬤⬤⬤⬤⬤⬤⬤⬤⬤⬤	105000	fresh

a Y. Ota, Earthrise Farms, California 1995
b *Diet for a Small Planet,* 1982, pg. 76-77, Dr. David Pimentel, Cornell University, 1981.

Fresh water is one of the world's most critical resources. Spirulina can use brackish or alkaline water, unsuitable for agriculture. Growing algae for food will become more attractive since it does not compete with needs for drinking or agriculture.

More efficient energy use.

Spirulina requires less energy input per kilo than soy, corn or beef, including solar and generated energy. Its energy efficiency (food energy output per kg / energy input per kg) is 3.5 times higher than soy, 1.4 times higher than corn, and over 100 times higher than grain fed beef. As cheap energy resources are depleted in the next 20 years, costs of energy dependent foods will rise with energy prices.

Energy Efficiency
(Million Kjoules Per Kilogram of Product)

	Total Energy Output	Food + Residual Energy Output	Energy Output/ Input
Spirulina[a] 65% protein	3.8[b]	23	6.1
Soybeans[b] 34% protein	11.7	13.8	1.2
Corn[b] 9% protein	5.5	16.5	3.0
Grain-fed Feedlot Beef[b] 20% protein	456	16	.04

[a] Y.Ota, Earthrise Farms, California 1995
[b] Leesley, et al. "A low energy method of manufacturing high-grade protein using spirulina Universitiy" of Texas, 1980, Pimentel, 1975. USDA

A big oxygen producer.

Forests help absorb carbon dioxide. Trees are the best land plants for fixing carbon, from 1 to 4 tons per hectare per year.[9] Spirulina is even more efficient. In the California desert, spirulina fixes 6.3 tons of carbon per hectare per year and produces 16.8 tons of oxygen. In the tropics it is 2.5 times more productive.[10]

Understanding carbon budgets is an emerging field of study. For example, how much new forest area should be planted to offset carbon emissions? In 1988, World Resources Institute recommended a U.S. utility plant 52 million new trees in Guatemala to offset CO2 emissions of a new U.S. coal burning power plant over 40 years.[11] In the 1990s, carbon budgets will be taken seriously as our planet struggles with global warming from the buildup of atmospheric CO2.

Three problems facing global food production
Agricultural: Limits of the green revolution.

Many consider the green revolution of the 1960s and 1970s to be a successful agricultural achievement. India and Indonesia are often cited as examples of new food self-sufficiency, with an average yield per hectare increasing from 1.1 to 2.6 tons. Yet, the Worldwatch Institute concludes this approach has failed in several ways:[12] Success has not been distributed evenly. New seeds, fertilizers and pesticides boosted yields of export crops of wealthier farmers with money and access to irrigation. Yields of locally consumed food of subsistence farmers on marginal rainfed land did not benefit as much.

By 2020, our planet will have 2 million more people, over 7 million total, with an unprecedented need for food. As people grow wealthier, especially across Asia, there is demand for more diverse foods such as meat and dairy which require large amounts of grains. Unfortunately, Worldwatch reports grain growing area peaked in 1981, and has fallen more than 5% since. Yields are down as irrigated area has begun to shrink and fertilizer use is being curtailed due to diminishing returns.[13] Big losses of fertile farmland continue to result from desertification and soil depletion. We are running out of space.

World Resources Institute agrees new approaches are needed. "Agricultural research has changed its perspective since the days of the Green Revolution. Although production oriented research which was responsible for developing high yield varieties is still at the forefront, some researchers are now increasing emphasis on the needs of poor farmers and on ecologically sustainable agriculture."[14]

Political: Food is not evenly distributed.

Many food experts claim the problem is not production, it is equitable distribution. A landmark book in the 1970s, Food First, claimed food is produced for profit, not people.[15] Although the world food supply is adequate to end hunger, chronic hunger persists, victimizing one third of the world's people, and primarily children.

Today, the developing world is even worse off with its huge debt burden. Wealthy elites of developing countries took on debt in the 1970s for over-ambitious and ill-conceived development schemes. Now these countries cannot repay the interest on the debt, and their people suffer with austerity programs imposed by the IMF, the World

Bank, and governments. These debtor countries are using their limited agricultural resources to grow export crops for hard currency to pay the interest on their debt, and not for food for their own people.

The underlying cause of political instability in these regions is the inability of people to control their own resources. Many would agree with the theme of the Hunger Project: "Although the world can feed itself, food is not evenly distributed. Hunger persists wherever people lack opportunity to participate in their society and end hunger."[16]

Environmental: Food growing area are declining.

There are few signs these inequities will soon change. Not only do many people continue to suffer chronic hunger, the environment continues to deteriorate. Desertification and soil depletion worsen from the stress of chemically produced export crops, livestock overgrazing, poor soil management, pollution and rainforest destruction.

A 1983 U.N. Food and Agriculture Organization (FAO) report concluded the developing world as a whole is capable of producing sufficient conventional foods to sustain its own population by the year 2000.[17] Yet, because the unrestricted movement of food within the developing world is unrealistic, 65 countries would have insufficient resources to meet their food needs by 2000. The excess population dependent on imported food will rise to 440 million people. These regions include over 20% of Africa where millions of people live.

Worldwatch reports desertification claims 15 million acres worldwide each year, an area the size of West Virginia. According to the U.N. Environmental Program (UNEP), "11 billion acres – 35% of the earth's land surface – are threatened by desertification and, with them, fully one-fifth of humanity."[18] Successful economies gobble up cropland. "Asia is losing ground. The development boom is eating away Asia's cropland. Unless this trend is reversed, the continent's leaders may find themselves facing a new national security problem: food."[19]

All three problems combine for maximum impact: 1) The green revolution has reached its limits, 2) world food distribution is not likely to change dramatically, and 3) cropland is lost due to economic growth and environmental deterioration. As farmed land turns into desert, or is paved over for infrastructure, less fertile land is available, while population is growing. To face these problems we need unconventional food sources with higher yields.

Advantages of spirulina production

This is where a new food resource such as spirulina can be beneficial. For food self-sufficiency, these areas need new food production and economic opportunity while simultaneously restoring the environment. Unconventional foods such as spirulina, microalgae, aquaculture, salt-tolerant crops and new drought-resistant grains and legumes offer help for these arid regions. New types of seawater-tolerant spirulina may become available. Genetically improved algae may have a higher content of desired nutrients or grow better in cold weather.

Facing a serious loss of cropland and higher food imports, China has declared spirulina a national food priority, In the last five years, over 50 spirulina farms have sprung up, mostly in southern China, making this country one of the largest producers.

Areas of chronic malnutrition are most common in the arid tropics and subtropics. Here, diets are high in carbohydrates and sugars, but low in protein, certain vitamins and minerals. Spirulina thrives best in these locations, and is a perfect complement to the typical diet. It speeds recovery from malnutrition. It is more digestible than other plant, dairy or meat products. This is important for victims of malnutrition with poor ability to digest food.

In a worst case scenario, global warming may create drastic climate change. The world's most productive crop regions could fail to produce food because of excessive heat and lack of rainfall. This drought situation is feared for the great North American grain belt. A climate crises could trigger worldwide food shortages and a scramble to find solutions. Microalgae require less land and water than other protein foods and can grow in hot climates where other crops cannot.

Today, while commercial farms grow spirulina as a health food for millions of people on six continents, lake harvesting and village-scale farms in Africa, Asia and South America can produce food for local people. Using appropriate technology, they address needs of waste treatment, soil and water quality, reforestation and food production, offering a context for the revival of the environment and the economy of villages and entire regions.

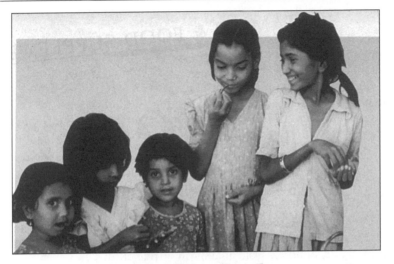

Chapter 8

Spirulina in the developing world

Each day, 40,000 children die of malnutrition and related diseases. One dream behind spirulina was developing it as a new food resource for a hungry world. One tablespoon a day offers remarkable health benefits.

The United Nations World Health Organization (WHO) in Geneva has confirmed:

"Spirulina represents an interesting food for multiple reasons, and it is able to be administered to children without any risk. We at WHO consider it a very suitable food."[1]

- Benefits of one tablespoon a day.
- Bioneering visions.
- Integrated village scale systems.
- Three experimental village projects.
- Child nutrition and small scale farms in India.
- Harvesting from lakes in Myanmar.
- Hope for the world's children.

Benefits of one tablespoon a day

Spirulina offers remarkable health benefits to an undernourished person. Its rich beta carotene can overcome eye problems caused by Vitamin A deficiency. The protein and B-vitamin complex makes a major nutritional improvement in an infant's diet. It's the only food source, except mother's milk, containing substantial amounts of an essential fatty acid, GLA, which helps regulate the entire hormone system.

One tablespoon a day can eliminate iron anemia, the most common mineral deficiency. Spirulina is the most digestible protein food, especially important for malnourished people whose intestines can no longer absorb nutrients effectively. Clinical studies have shown it helps rebuild healthy intestinal flora.

These health benefits have made it an excellent food for rapid recovery of children from malnutrition related diseases in Mexico, Togo, Romania, China, Rwanda, Zaire, India, Ukraine and Belarus.

Bioneering visions

In the 1960s and 1970s, small groups of scientists, promoters and visionaries understood what this microalgae could become. Larry Switzer, wrote about his hope for a breakthrough in food production:

"It had to be more productive than conventional agriculture ... adaptable to different climates and cultures ... appropriate ecologically, economically and socially ... independent of the vested interests in world food production and distribution ... capable of relying on renewable energy and waste or abundant raw material resources. It would have to represent a major expansion of the photosynthetic energy base that supports all life on Earth."

8.2. Larry Switzer, 1993.

"Finally, it would have to radically improve the supply, distribution and consumption of essential protein to millions of pregnant and nursing mothers, infants and children. It is absolutely critical to provide nutrition to the deprived embryos and infants of the world to preserve the precious creative genius that is waiting to be released from each fully developed human mind."[2]

In 1976, Switzer proposed to the Nigerian Government that oil money finance farms along the shores of Lake Chad to feed the country's growing urban masses. But the novelty of algae, lack of interest in food production, and a political coup doomed this plan. The lesson learned was to develop the technology and a consumer market in the United States first, and then apply it to the Third World.

Switzer founded Proteus Corporation and the Earthrise Company which built the first farm in California, the forerunner to Earthrise Farms. Earthrise introduced spirulina in 1979.

Beginning of village scale technology

Parallel to commercial production has been the notable work of Ripley and Denise Fox. Over the past 20 years, they developed the Integrated Health and Energy System and founded the non-profit organization, *Association Pour Combattre la Malnutrition par Algoculture* (ACMA) to help fund these projects.[3]

The Foxes grappled with the most fundamental problem: the lack of economic opportunity. Large commercial farms could produce spirulina, but if the hungriest people had no money to buy it, they would never get any. They must grow it themselves, but without the money to buy nutrients, how could a village project be self-sustaining?

Villages have an abundance of human and animal wastes – a source of pollution and disease. Intestinal parasites are spread by contact with wastes. These parasites consume 30% of the food eaten by people. The first step in improving health and increasing effective food production is eliminating intestinal parasites through sanitation and waste treatment. Properly handled, wastes can be converted to energy, compost, clean nutrients and even food. "One simply recycles the wastes already present in the village," says Dr. Fox. This is the foundation of the Integrated Health and Energy System, described in his books, *Algoculture: Spirulina, Hope for a Hungry World,* and *Spirulina, Production and Potential (1996).*[4]

The integrated health and energy system

The enthusiasm, participation and training of the village people are vital to the success of the integrated system. Villagers must be shown how they can improve their own sanitation, health, nutrition and village ecology, and the means to generate income from the compost, biogas fuel, fish and algae. The system is flexible enough to accommodate various cultural traditions.

Latrines eliminate the source of intestinal parasites. Where houses are spread out over a wide area, family latrines can produce compost to improve soil fertility. In a village center or market, central latrines can provide continuous wastes to a biogas digester.

Animal manure and plant waste are added to the digester which ferments the waste and breaks it down into gas, liquids and solids. A simple gas separator separates biogas into methane (a fuel for cooking and lighting) and carbon dioxide (a nutrient for spirulina). The liquid effluent is sterilized in a series of solar heated pipes and becomes a safe mineral nutrient source for the spirulina pond. Sludge is removed from the digester and composted before spreading on soils.

Solar photovoltaic panels produce electricity to run the paddlewheel in the pond, auxiliary lighting for other facilities, as well as recharging batteries. The spirulina is screened from the pond water and can be added to the aquaculture pond to feed fish or dried in a small solar drier for human food.

Other important elements of the system include a clean water system and tree planting. Livestock pens are strongly encouraged to help prevent overgrazing, the primary cause of desertification in Africa, leaving whole regions denuded and vulnerable to drought.

This Integrated Health and Energy System is a systemic approach to community problems. To succeed, the people themselves need to participate and alter some of their old habits to help revitalize their community. Each successful project becomes the inspiration and training center for the surrounding villages.

The design for the Integrated Health and Energy System won the prestigious 1987 *European Award for Appropriate Environmental Technology*, sponsored by the European Economic Community and the United Nations Environmental Program.

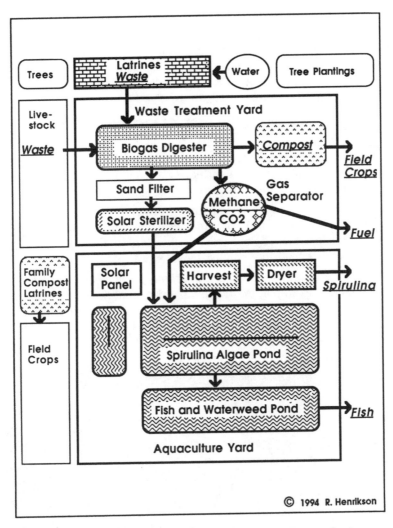

8.4. Layout of the Integrated Village Health and Energy System.

Three experimental projects

"Every nation is supported on the shoulders of its villages," says Fox. "We believe by providing technical assistance to improve sanitation and agricultural output and by saving the trees, we can increase the vitality of these villages."[5] Since 1981, Ripley and Denise Fox have worked to bring these tools to the Third World.

Karla, India – the first integrated system

Karla village is near Wardha, famous for Mahatma Gandhi's ashram. The Center of Science for Villages (CSV), an organization inspired by Gandhi's vision that India's strength lies in the strength of its villages, operates the village system in Karla.

Today, this project derives income from compost and fish sold to local villagers, and from spirulina sold in Bombay. The CSV has distributed spirulina cookies and noodles to local children with great success, and chapatis are another local favorite.

Since algae grown on waste nutrients must be proven safe to consume, it was important to show pathogens in the original human and animal waste were destroyed in the biogas digester and solar sterilizer. A six month study showed pathogens were virtually eliminated and were lower than in other local foods.

8.5. Dr. Ripley Fox and assistants discussing cultivation in Karla, India.

Farende, Togo – a remote African village

Less than 100 years ago, Northern Togo in West Africa was home to lush forests, elephants and tigers. Today the big animals are gone and little stands in the way of the Sahara sweeping south. These arid grasslands are home to an exploding human population. Family compounds dot the fields and hillsides. The hard working Kabyé people grow millet and root crops. But soil quality in marginal. Growing numbers of people stress the carrying capacity of the environment. Severe problems lie ahead.

The remote village of Farende participated in an experimental project to grow spirulina. Solar panels charge truck batteries that drive the paddlewheels in the basins. A small 100 square meter basin can grow enough to supplement the diet of 100 children a day. Pouring the pond water through a screen, spirulina becomes a thick paste, and its loaded on a screen in a solar heated dryer. Dried algae is distributed at the health clinic.

**8.6. Tending the growing basin in Farende, Togo, 1989.
Solar dryer in background.**

Undernourished children take spirulina as a daily supplement at the health clinic. The head nurse tells the mothers about its benefits. One tablespoon a day mixed with water brings remarkable results.

8.7. Dr. Ripley Fox and assistants harvesting through screens.

Children find the taste of this "green medicine" acceptable and within a week begin to show signs of health improvement and gain weight. Mothers from the surrounding countryside brought their children every week to participate in a clinical feeding study in 1989.

8.8. The local health clinic nurse introduces children to spirulina.

8.9. The people of Farende, Togo. (Photocollage by R. Henrikson).

From 1984 to 1990, the Foxes assisted the Eglise Evangelique du Togo in building this experimental system, operated by two young village men appointed by the elders. Also cooperating was the U.S. Peace Corps, teaching people about soil composting, vegetable gardening, conservation and growing trees. The first corporate sponsors were DIC of Japan and Earthrise Company of California.

San Clemente, Peru – an urban shantytown

The hot arid coast of Peru with its poor land and scarce fresh water is a typical climate for spirulina. San Clemente is a new town near Pisco, its population swelling from 10,000 to 40,000 in less than five years. The town, perpetually broke, did not have sufficient funds for a sanitation or water system for the shantytown.

This integrated system was funded by ACMA, private donations from France, and the Earthrise Company. The mayor and towns people, the Cooperacion Popular and members of the French embassy opened the project in 1987. Plans were made to distribute algae through the local canteens to slum children whose parents could not feed them properly. Unfortunately, due to the political chaos, civil war and local unrest, work was halted at the close of 1988.

Reducing eye disease in India

The world's largest spirulina nutrition program with 5,000 children was finished in 1992. Children near Madras consumed one gram a day for 150 days. This small amount provided the daily requirement of beta carotene (Vitamin A) which helps prevent blindness and eye disease. A symptom of Vitamin A deficiency, *Bitot's spot,* a scarring of the conjunctiva of the eye, decreased from 80% to 10%.[6]

Spirulina was given to children in extruded noodles, sweetened with sugar to preserve beta carotene. Called *"Spiru-Om,"* it was well accepted by the children. This project was sponsored by the Indian government and was lead by Dr. C.V. Seshadri of the Murugappa Chettiar Research Center in Madras. It is hoped this program can be extended to other districts in India.

8.10. Dr. C.V. Seshadri, with Dr. Ripley Fox and R. Henrikson, 1993.

One goal of the program was to provide an alternative to the current Vitamin A therapy, giving massive doses of imported pure Vitamin A every six months to children.

8-11. A spoonful of "Spiru-Om" noodles is given to a small child.
(Courtesy of C.V. Seshadri[6])

Family scale cultivation in India

An even better solution is growing spirulina locally, and giving it to children. Another government sponsored project in Southern India, provides small backyard basins to women for family nutrition.[7] This could develop into local village networks of ponds to combat Vitamin A and general immune deficiency conditions.

8.12. Home cultivation and harvesting in a village in Tamil Nadu, India. (Courtesy of C.V. Seshadri[6])

The *All India Coordinated Project on Algae* began in 1976 to harness algae as biofertilizer and to cultivate it for human nutrition and animal feed at both commercial and rural levels. Original work began with scenedesmus, green algae, but later, spirulina was chosen because of its advantages. In 1991, the Indian government launched large scale nutritional studies. To demonstrate national interest, the Ministry of Health issued official standards for food grade spirulina.

India may be the only country in the world conducting a joint effort by many government agencies covering all aspects of spirulina, from simple cultivation basins to large scale commercial farms. The government has sponsored large scale nutrition studies with animals and humans and has investigated therapeutic uses.[8]

Harvesting from lakes in Myanmar

Four spirulina lakes were identified and studied beginning in 1984. First production began in 1988, and by 1993, capacity was about 30 tons per year. During the peak or blooming season around March, people in boats collect a dense cream-like concentration of spirulina in buckets. Although this bloom lasts for only one month, it represents 70% of total production. During the rest of the year when the concentration of algae in the lakes is low, lake water is filtered through fine polyester cloth on inclined screens.

8.13. Harvesting with buckets from a lake in Myanmar, 1992.
8.14. Spirulina cream poured into cloth cones for dewatering.
(Courtesy of Min Thein[9])

Harvested cream is poured into cloth bags, washed with fresh water and placed under pressure. This paste is extruded into noodle like filaments and dried in the sun. The sun dried chips are taken to a factory, pasteurized to treat bacteria, and pressed into tablets.

In the past four years, over one million bottles of locally grown tablets have been sold in Myanmar. People report improvements in the intelligence of children, fertility of couples, faster wound healing, increased mental awareness, more energy for older people, and better meditation practices for monks and nuns.[9]

In the future, by building growing ponds along the four lakes and installing a more efficient harvest and drying system, Myanmar has the ability to become a major producer.

Hope for the world's children

In the developed world, spirulina is an established health food with a growing reputation of making healthy people even healthier. It has been recognized as a "national food" by the two most populous countries, India and China, where cropland is in short supply. Both countries are becoming major world producers.

In India and Vietnam, it is recommended for nursing mothers. For newborns, malnutrition is often caused by a lack of mother's milk, the mother herself often being ill. Spirulina given to the mother helps a return to lactation and the babies rapidly gain weight.[10]

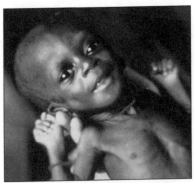 

6.15-16. Baby saved from malnutrition by spirulina, before and after in Togo, West Africa.

In Ho Chi Minh City, Vietnam, in 1990, doctors conducted nutrition trials in two orphanages. Food formulas containing 5% spirulina had a better effect than soya at much lower levels.[11]

In a bush hospital in Zaire in 1990, spirulina in corn flower cookies improved health of children with severe protein-energy malnutrition.[12] In the Central African Republic in 1992, a clinic treated 200 children a day, improving the health of those with kwashiorkor and marasmus.[13] In Rwanda in 1993, children with kwashiorkor were given algae at a dispensary. After 15 days, their mothers wanted to buy spirulina and learn how to grow it in their village.[14]

Mothers and children around the world are embracing spirulina. In this coming decade we hope to witness an even more rapid acceptance to benefit the next generation.

Chapter 9

Microalgae's role in restoring our planet

Understanding the role of microscopic algae, the foundation of life, can help us develop restorative models of personal and planetary health.

Microalgae is an essential part of Earth's self-regulating life support system. Restoration is the next big growth industry. Innovative schemes and dreams using microalgae promise to help regreen the desert, refertilize depleted soils, farm the oceans and encourage biodiversity.

- Algae life support systems in space exploration.
- Terraforming Mars with blue-green algae.
- Ecological communities on Earth.
- Farming natural alkaline lakes.
- Schemes for regreening the desert.
- Giant seawater farms.
- Algae biofertilizers restore soil fertility.
- Growing ocean algae to reduce global warning.
- Planetary restoration and individual transformation.

Spirulina is going to outer space

The U.S. National Aeronautics and Space Administration (NASA) plans to put a space station into orbit soon. Critical to the success of a permanent manned space station is a regenerative life support system called CELSS. NASA contracted scientists to develop a system for providing oxygen and food for humans living in space for long periods. Because the first small space stations will have little room for rows of plants, scientists are looking at algae.

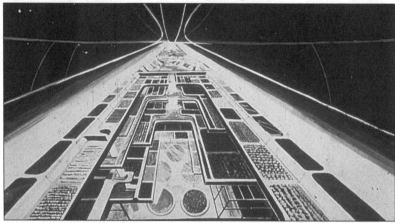

9.2. Artists conception of a space colony, depicting fields and ponds.

Early experiments demonstrated that shrimp and mice could live in a completely sealed environment with a food supply. Algae consume carbon dioxide exhaled by the shrimp and mice, and exhale oxygen for them to breathe. These experiments proved algae can make enough oxygen to keep animals alive in a small, closed system.[1]

Algae have a higher photosynthetic efficiency, releasing more oxygen and producing more food than any other plant. Nutrients will come from the carbon dioxide exhaled by humans and recycled human and food wastes. This solves the problem of disposing of wastes during space travel. Rapid growing algae turn waste into purified water, nutritious food and oxygen sufficient to support humans.

NASA-Ames Laboratory and the National Aerospace Lab in Japan proposed a food production system using spirulina and chlorella in a photosynthetic gas exchanger[2]. Wastes are heated to very high

146

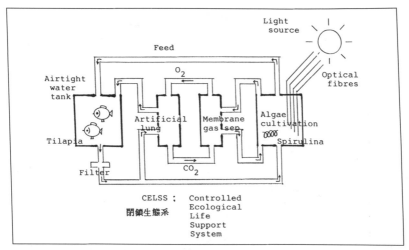

Light source

Feed

O_2

Airtight water tank

Optical fibres

Artificial lung

Membrane gas sep.

Algae cultivation

Tilapia

Spirulina

Filter

CO_2

CELSS : Controlled Ecological Life Support System

閉鎖生態系

9.3. Diagram of a semi-closed microalgae gas exchange experiment by the National Aerospace Laboratory of Japan.

temperatures, turning them from semi-solids to gases. These gas nutrients are pumped into the algae tank along with carbon dioxide exhaled by the astronauts. As the algae grows, it can be continuously harvested and made into various food forms. CELSS includes a special computer system for controlling this gas exchange system.[3]

Kennedy Space Center used spirulina from Earthrise Farms to research growing fish in space stations. More ambitious NASA plans include a larger self-sufficient system to grow enough algae and plant crops to permanently support human beings. Special lenses and optical filters would collect sunlight in space.

Terraforming Mars with blue-green algae

Life on Mars was announced in 1996 – or at least it may have existed once. A dozen space probes are heading there in the next decade. Scientists are talking of colonizing the red planet. They say it may be possible to terraform Mars – make it more like Earth so we can live there. "The recipe is simple. Add nitrogen and oxygen to the atmosphere; pump water to the surface; cook for decades, spicing first with cyanobacteria, then with all the rest of Earth's plants and animals, adding them in the order they evolved here ... Terraforming Mars would take 300 years at least."[4]

147

Ecological communities on Earth

Closed ecological systems need the efficiencies of microalgae living at the base of the food chain. Whether it is the closed atmosphere in a space station or the ecological cycle of a third world village, both are regenerative models based on our planet's living ecosystem.

Solar powered communities of the future may incorporate the benefits of microalgae. These ecological communities would be designed for high agricultural productivity, simultaneously restoring the surrounding environment. Communities envisioned in the book *Bioshelters, Ocean Arks and City Farming* are coming, assert the authors, Nancy and John Todd. "New biotechnologies, information, and biological components are being assembled into ecosystems capable of providing a diversity of foods in relatively small spaces."[5]

Bioshelters and city farming will use information technology to become viable alternatives to capital and oil dependent agriculture. Both may incorporate algae production along with intensive organic gardens and aquaculture. On a small area, productivity can be optimized, freeing up croplands for common areas or forests. "An ultimate goal might be that for every acre which is farmed another would be set free."[6]

In 20 years, new communities will provide an attractive alternative to land consuming suburbia or the inner city asphalt jungle. Combining these communities with algae microfarms for growing food would speed regreening our blue-green planet.

Family scale cultivation

Many people have asked how they can grow spirulina for their family or community in their own back yard. Although in India, low technology methods are used, nevertheless, successful small scale cultivation poses a real challenge. A recent book, *Spirulina: Production & Potential*,[7] by Dr. Ripley Fox, reviews the knowledge, equipment and funds needed for family scale cultivation: conditions for growth, nutrient media, growing basins, essential laboratory equipment, starting the culture, harvesting, drying and storage. The chapter on troubleshooting is 27 pages long!

A new resource is being published in French by Jean Paul Jourdan, explaining how to cultivate spirulina on a family scale: *Cultivez Votre Spiruline – Manuel de Culture Artisanale.*[8]

Farming natural alkaline lakes

Spirulina lakes are found in Peru, Chile, Burma, Australia and stretch across the Sahara and East Africa, near millions of chronically undernourished people. For 20 years, scientists and visionaries have proposed harvesting algae from the lakes in Ethiopia and Kenya.

Assessing the environmental impact is the first stage. A pioneer in spirulina farm design, lake ecologist Alan Jassby believes harvesting wild spirulina is too simplistic an answer, since the natural growth is unpredictable, the density is too low for efficient harvesting, and continual harvesting could exhaust nutrient sources in the lakes. Half of the world's flamingo population live in these lakes on spirulina. They are a tourist attraction, contribute to the ecological cycles of these lakes, and should be protected. Harvest of wild algae must be checked for contamination by toxic algae or other natural or man-made pollutants that could pose a public health hazard.[9]

The best approach is building cultivation ponds beside these lakes. Without disturbing the larger ecosystem, these farms can share the resources with the wilderness in a sustainable fashion.

How to distribute the food produced to poor people with no money to buy it? If lake farms are merely set up as commercial ventures, then most spirulina would be exported to raise hard currency. A project cooperatively sponsored by governments, international agencies or business, might be able to supply needy indigenous people.

In Central and East Africa, 20% of the people may be infected with the AIDS virus. Here, in the center of this problem is the solution: huge lakes filled with blue-green algae. If sulfolipids are shown to stop the AIDS virus, an international effort against AIDS could mobilize resources to develop lake cultivation, extracting sulfolipids for an anti-AIDS drug. The algal byproduct would be 65% protein, rich in vitamins and minerals – still the most nutritious food.

Farming East African lakes could create major opportunity: 1) as the source of an effective AIDS drug for Africans, 2) for export to the West for hard currency, 3) as the source of a new food to help feed Africa, and 4) to relieve environmental pressure on the other food growing areas. These lakes could represent an economic miracle for East Africa.

Schemes for greening desert coastlines

Large areas cannot be reforested if millions of people surrounding them cut down their forests for food and fuel. Economic opportunity zones alongside reforestation zones will be critical to the success of reforestation. In areas with depleted natural resources, microalgae can create opportunity with a fraction of the land and water as conventional crops. Several restoration projects have been proposed.

Once scientists learn to successfully cultivate microalgae in seawater, new food growing areas can use the more than 10,000 miles of accessible desert coastline in hot climates: Mexico, Peru, Chile, West, North and East Africa, Egypt and the Arabian peninsula and India.

Giant seawater farms

In *Spirulina, Production & Potential,* Dr. Ripley Fox envisions building huge seawater farms along desert coastlines.[10] He proposes a network of 25 farms, 120 hectares each, providing 10 grams a day for 30 million children, the number at high risk of dying from malnutrition and related diseases. He claims the construction and operating cost would be less than one day of the 100 day Gulf War.

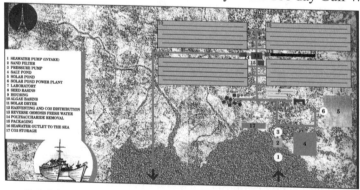

1 SEAWATER PUMP (INTAKE)
2 SAND FILTER
3 PRESSURE PUMP
4 SALT POND
5 SOLAR POND
6 SOLAR POND POWER PLANT
7 LABORATORY
8 SEED BASINS
9 HOUSING
10 ALGAE BASINS
11 SOLAR DRYER
12 HARVESTING AND CO2 DISTRIBUTION
13 REVERSE OSMOSIS FRESH WATER
14 POLYSACCHARIDE REMOVAL
15 PACKAGING
16 SEAWATER OUTLET TO THE SEA
17 CO2 STORAGE

9.4. Layout of giant seawater farm (courtesy Dr. Ripley Fox).

Each farm would have twelve 10 hectare ponds lined with plastic film to make them watertight. Floating paddlewheels circulating the ponds would be powered by nearby solar ponds producing electrical energy, a technology pioneered in Israel. The main nutrient, carbon dioxide gas, would be recovered from fuel-fired power, chemical and heavy industries that normally release into the atmosphere.

Governments would encourage industries to collect CO_2 in pressurized tanks using tax incentives, and then transport those tanks on unused military transport ships to the algae farmsites. Drying would be accomplished by huge solar drying tubes, 2 meters in diameter and almost 500 meters long.

Dr. Fox suggests these farms be owned by international, humanitarian non-governmental organizations, and financed by governments by monies destined for military budgets. A network of distribution centers in malnutrition zones would distribute the dried algae.

Algae biofertilizers restore soil fertility

Some of these schemes may be able to use the magic of nitrogen fixing blue-green algae cultivated in Southeast Asian rice paddies for a thousand years to increase rice production. A 1981 U.N. FAO report *Blue-Green Algae in Rice Production* documented the possibilities of blue-green algae replacing chemical fertilizers and rebuilding the structure of depleted soils.[11]

In India, blue-green algae is grown in shallow earth ponds. When the water evaporates, the dried algae is scooped up and sold to rice farmers. This natural nitrogen source is only one-third the cost of chemical fertilizer and it increased annual rice yield in India and several other countries an average of 22%. Where chemical fertilizers are not used, algae gives the same benefit as 25 to 30 kg of chemical nitrogen fertilizer per acre. Where chemicals are used, algae reduces the chemical dose by the same amount.

Algae can also increase plant growth, such as the green algae clamadamonas, rich in polysaccharides, which help recondition soil fertility and build soil structure to retain more moisture. Algae have plant growth regulators, and by inoculating soil with algae, plant productivity can be enhanced. Scientists are looking at DNA engineering of blue-green algae to improve their nitrogen-fixing efficiency.[12]

Algae biofertilizers can be used for rice paddy cultivation. They can inoculate soils to increase food productivity along desert coastlines, near alkaline lakes and in villages. Used as biofertilizers, algae offer yet another way to help feed people through soil renewal, providing economic opportunity without resorting to the vicious cycle of chemical fertilizers, soil exhaustion, and dependence on imports.

Can growing ocean algae reduce global warming?

Most scientists agree several actions can halt the buildup of greenhouse gases in the atmosphere: 1) stop adding chlorofluorcarbons (CFCs) to the atmosphere to halt ozone layer depletion and global warming; 2) restrict methane and carbon dioxide emissions which contribute to global warming; 3) stop destroying forests and coral reefs because their stored carbon is released into the atmosphere; 4) limit human and livestock populations that stress the carrying capacity of the natural environment; and 5) plant forests to remove carbon from the atmosphere to be stored in trees and to release oxygen.

A novel idea is raising ocean productivity to remove massive amounts of carbon dioxide from the atmosphere. Phytoplankton absorb atmospheric carbon dioxide. Some scientists proposed enriching the ocean with iron particles. This would stimulate algae blooms and photosynthetic plant growth which fixes carbon in organic forms, leading to eventual storage as minerals in coral reefs.[13]

In a dramatic experiment in 1996, researchers spread 1000 pounds of iron particles across nearly 30 square miles of ocean near the Galapagos Islands. "In a single week we created a new world in the middle of the ocean. We turned a patch of clear blue ocean water into a big green hayfield, just like going from a desert to forest." The plankton consumed more than 4 millions pounds of carbon from the atmosphere.[14] However, scientists remained skeptical of dumping iron all across the world's oceans because of unpredictable effects on global ecology and because it might still prove ineffective at cleaning enough carbon dioxide to slow down global warning significantly.

The challenge of restoration:
Ahead to the beginning

We are beginning to discover the ways we can work with the original photosynthetic life form to restore this planet – limited only by our imagination As this new vision of the world unfolds, the few ideas mentioned here will blossom into many more.

Life in the Earth's biosphere depends on the balance of gases in the atmosphere, as James Lovelock described in his Gaia theory. Lovelock believes how we grow food has the most profound impact on planetary ecological deterioration.

"Bad farming is probably the greatest threat to Gaia's health. We use close to 75 percent of the fertile land of the temperate and tropical regions for agriculture. To my mind, this is the largest and most irreversible geophysical change that we have made ... Could we use the land to feed us and yet sustain its climatic and geophysical roles? Could trees provide us with our needs and still serve to keep the tropics wet with rain? Could our crops serve to pump carbon dioxide as well as the natural ecosystems they replace? It should be possible but not without a drastic change of heart and habits."[15]

More and more people realize they can affect global food patterns by changing their own habits. Eating lower on the food chain, they eat less meat, more organic vegetables and grains and perhaps even algae. They are healthier, they help reduce environmental damage, and their choices can help return cropland and grazing land back to new forests.

We can consider what values and attitudes we express. These values result in economic choices that affect the world. We can review our personal choices about energy, water and land conservation, garbage and toxic products, spending and investing money in socially responsible business, and supporting environmentally sound politics. Changing these patterns, magnified by millions, can change the world.

Choices are making a difference already

Consumer purchasing decisions make a difference. Studies show people will pay more for natural and environmentally friendly food. Green business is becoming good business. A recent study on consumer purchasing decisions found: "One-third of Americans say that after price and quality, a company's socially responsible business practices are one of the most important factors in deciding whether or not to buy a brand. In fact, social responsibility was slightly more influential than advertising. This signals a change in public opinion."[16]

People do understand the connection between their personal health and the planet's health, and they are making choices to help restore our health.

As you think globally and act locally,
consider the role of microalgae like spirulina,
coming into its own as a world resource.
The oldest photosynthetic life form is back.
It represents a return to the origins of life –
for a new Earthrise.

153

Procession

Spirulina speaks to the human species, on behalf of the first species – algae.

"We, the oldest algae, began photosynthesis at our greatest moment in evolution to initiate the unfolding of life. We can assist you, for we have a fundamental role to play.

Sometime in the future, you will look back at these next 20 years. The transformation of your attitude toward your body, your species and your planet will seem to have been remarkably rapid. This wonderful adventure will be the most magnificent and creative moment of your evolution in becoming fully human.

Understanding your evolution is so simple yet so profound. In realizing what harm you have done to our planet will come your liberation from the past and your greatest gift to life. You will understand your role is not to control, but to participate with nature.

We are all composed of the same elementary particles dating from the beginning of creation. Your human form is yet another alignment of particles which dance together with all particles. Rejoice in your connection with all life.

In rediscovering your relationship with the natural world, you will heal yourself and our planet. We are excited for your opportunity in these next 20 years. Now go forth and express yourselves and participate fully in the beauty, the wonder and the glory of our unfolding creation."

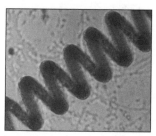

Appendix A
Quality and safety standards

Spirulina has been marketed and consumed as a human food and has been approved as a food for human consumption by many governments, health agencies and associations of these countries:

Argentina	Australia	Austria
Bahrain	Bahamas	Bangladesh
Belarus	Belgium	Brazil
Bulgaria	Canada	Chad
Chile	China	Colombia
Costa Rica	Croatia	Czech Republic
Denmark	Ecuador	Egypt
Ethiopia	Finland	France
Germany	Greece	Guam
Gulf States	Haiti	Hong Kong
Hungary	India	Iceland
Indonesia	Ireland	Israel
Italy	Jamaica	Japan
Kenya	Korea	Kuwait
Liechtenstein	Luxembourg	Macedonia
Malaysia	Mexico	Myanmar
Monaco	Netherlands	New Zealand
Nigeria	Norway	Peru
Philippines	Poland	Portugal
Romania	Russia	Saudi Arabia
Singapore	Slovenia	South Africa
Spain	Sweden	Switzerland
Taiwan	Thailand	Togo
Turkey	Ukraine	United Kingdom
United States	Venezuela	Vietnam
Yugoslavia	Zaire	Zimbabwe

A.1. Spirulina food safety guidelines used by US and Japan growers

Criteria	USA[a]	Japan[b]
Moisture	< 7 %	< 7 %
Bacteriological:		
Standard plate count	<200,000/ g	<200,000/ g
Mold	< 100 / g	< 100 / g
Yeast	< 40 / g	< 40 / g
Coliforms	neg.	neg.
Salmonella	neg.	neg.
Staphylococcus	neg.	neg.
Heavy Metals:		
Lead	< 1.0 ppm	< 1.0 ppm
Arsenic	< 1.0 ppm	< 1.0 ppm
Cadmium	< 0.05 ppm	< 0.05 ppm
Mercury	< 0.05 ppm	< 0.05 ppm
Insect fragments	< 30 / 10 g[c]	*
pesticides	neg.	neg.
herbicides	neg.	neg.
additives	neg.	neg.
preservatives	neg.	neg.
dyes	neg.	neg.
stabilizers	neg.	neg.
artificial ingredients	neg.	neg.
fillers	neg.	neg.

a. Published by Earthrise Farms, 1995.

b. Published by Dainippon Ink & Chemicals, Inc., Japan.

c. U.S. Food and Drug Administration guideline.

definitions:
<   = less than
*   = no set standard
/ g = per gram
neg. = negative

Food safety research

Spirulina has a history of use in Chad where locals traditionally consume 9-13 grams per meal, and these meals are from 10 to 60% of the meals.[1] "The attention of the U.N. FAO was attracted by the fact that algae was being consumed by humans. The FAO organized an educational campaign in Chad to encourage consumption of spirulina harvested from natural sources. More than 6000 meals were distributed under the supervision of the FAO and the campaign was crowned as a success. The program was suspended due to the outbreak of war."[2]

Another report stated "*dihé* (spirulina sauce) was served at the school canteen. One must admit the introduction of this product in the young people's food gave no problem in this region where the majority were Kanembou. But equally at Fort Lamy (now Ndjemena) we noted the product was accepted by other people."[3]

As previously reported in Chapters 4 and 8, spirulina was given to malnourished children and adults in clinical studies beginning in the early 1970s. Since the late 1970s, millions of people in the developed countries have used it as a health food supplement, taking 3 to 20 grams a day. Rarely are there any reports of allergies or sensitivities.

In the 1970s, spirulina underwent extensive safety studies with animals and fish. Independent feeding tests in France, Mexico and Japan showed no undesirable results and no toxic side effects on humans, rats, pigs, chickens, fish and oysters. Many independent rat feeding trials were conducted in Japan and no negative effects at all were found for acute or chronic toxicity or reproduction.[4,5]

In 1980, one of the most important and comprehensive animal studies was sponsored by the U.N. Industrial Development Organization (UNIDO) on rats and mice. Spirulina comprised 10% to 35% of the total diet. No second or third generation reproduction, fertility, lactation or birth defect problems were found. No cancer causing properties were found. No problems with heavy metals, nucleic acids, pesticides or bacteria were found. The study concluded any further research would demonstrate its complete safety as a human food.[6]

Toxicology research has continued through the 1980s and 1990s, showing spirulina has no peri- and postnatal toxicity in rats, no adverse effects on reproduction, including male and female fertility, and duration of gestation, and no increase in number of abnormal offspring.[7,8,9]

A.2.	Spirulina food standards

**microbiological quality requirements
of France, Sweden, Japan and Earthrise Farms (USA)**

Standard	*France[a]*	*Sweden[b]*	*Japan[c]*	*E Farms[d]*
Moisture	*	*	<7 %	<7 %
Standard Plate Count	<100,000/g	1,000,000/g	<200,000/g	<200,000/g
Mold	*	<1000/g	*	<100/g
Yeast	*	*	*	< 40/g
Coliform	<10/g	<100/g	neg.	neg.
Salmonella	neg.	neg.	*	neg.
Staph	<100/g	<100/g	*	neg.

a. Superior Public Hygiene Council of France, 1984, 1986.
b. Ministry of Health, Sweden.
c. Japan Health Foods Association, auth. by Ministry of Health and Welfare.
d. Earthrise Farms, 1995.

definitions:
< = less than   * = no set standard
/g = per gram   neg. = negative

Nucleic acid safety research

Spirulina has about 4% nucleic acids (DNA and RNA), lower than chlorella and other microalgae, yeast and fungi (6-11%). Although there was once some concern that eating microalgae might increase uric acid levels because of the nucleic acids, there is little evidence to support this. In fact, one study found that uric acid levels did not increase in humans taking up to 30 grams a day of chlorella protein (50 grams of chlorella).[10] Since spirulina is lower in nucleic acid content, eating up to 50 grams a day is safe as well, and means it can be safely used as major protein source.[11]

Published studies from independent laboratories around the world confirm the absence of any toxic effects even when it provides a significant amount of dietary protein.[12,13,14] Since its introduction as a human food in 1979, its success has confirmed the work of the earlier animal studies. Spirulina has been safely consumed by millions of people in North and South America, Asia, Europe and Africa.

Heavy metal safety research

Mercury, lead, cadmium and arsenic are widespread in our environment from ever-present industrial pollution. Heavy metals are toxic to humans in small amounts, and prolonged eating of foods contaminated with heavy metals can lead to long term health problems. Yet few companies or organizations disclose levels of these heavy metals in foods. One grower, Earthrise Farms, has published strict standards for heavy metals in spirulina.[15] A five year testing program in California showed heavy metals were either not detectable or extremely low. Based on 120 independent laboratory tests, Earthrise Farms set up some of the toughest standards for heavy metals.

Mercury was not detectable in 40 tests, and the standard for mercury was set at less than 0.05 parts per million (ppm). In comparison, the US FDA standard in 'aquatic animals' is 1.0 ppm, permitting over 20 times more mercury. Standards were set for cadmium (less than 0.05 ppm), lead (less than 1.0 ppm), and arsenic (less than 1.0 ppm). By comparison, the UN Protein Advisory Group standard for single cell protein permits higher heavy metals: 1.0 ppm for mercury; 1.0 ppm for cadmium, 5.0 ppm for lead; and 2.0 ppm for arsenic.

Algal toxin safety research

An important quality control issue surrounding production of blue-green algae (cyanobacteria) is the possibility of inadvertently harvesting other blue-green algae containing cyanotoxins. This is a risk when harvesting algae from natural bodies of water with mixed cultures of microscopic algae. Algal toxins are capable of causing widespread poisoning of animals and humans.[16]

In 1995-96, a group of leading microalgae producers sponsored research conducted by algal toxicologists. The result was a *Technical Booklet for the Microalgae Biomass Industry* as a guide to the use of a very sensitive enzyme linked immunosorbant assay (ELISA) and a protein phosphate inhibition assay (PPIA) for the detection of toxic microcystins and nodularins. These methods can detect, monitor and control cyanotoxins, so producers can assure a safe, nutritious product for human and animal food supplements.[17]

A.3.   Heavy Metal Guidelines

	Lead	Mercury	Cadmium	Arsenic
		< (less than) parts per million (ppm)		
Aquatic Animals				
(U.S. Food and Drug Admin.)	-	<1.0	-	-
Single Cell Protein				
(UN Protein Advisory Group)	<5.0	<0.1	<1.0	<2.0
Chlorella				
(Japan Health Foods Assn)	total heavy metals: < 20.0 ppm			
	-	-	-	<2.0
Spirulina				
(Japan Health Foods Assn)	total heavy metals: < 20.0 ppm			
	-	-	-	<2.0
Spirulina				
(Earthrise Farms guidelines)	total heavy metals: < 2.1 ppm			
(*undetectable)	<1.0	*<0.05	*<0.05	<1.0

A.4.   Tolerable amount per day

	Lead	Mercury	Cadmium	Arsenic
		mcg (micrograms) per day		
UN World Health Organization				
/ Food & Agriculture Org.	500	50	66-83	-
Earthrise Spirulina (10 gm per day)				
mcg of heavy metals:	<10	<0.5	<0.5	<10
percent of UN Guidelines:	<2%	<1%	<1%	-

A.5. Quality and Safety Standards for Spirulina
for the USA Natural Foods Industry

The Natural Products Quality Assurance Alliance (NPQAA)
and The Natural Nutritional Foods Association (NNFA)

Definition (dried powder as produced)

Spirulina is a blue-green microalga. Spirulina is cultivated in specially designed artificial ponds, harvested, dried and packaged in accordance with Good Manufacturing Practices (GMP), and quality controlled at each stage of the production process.

Legal Definition

According to the U.S. FDA (Talk Paper 6/23/82): as a food, spirulina can be legally marketed as long as it is labeled accurately and contains no contaminated or adulterated substances.

Spirulina powder as produced

Physical Appearance and Identity for spray-dried spirulina.

Spirulina should be readily identified as *Spirulina (Arthrospira) sp.* under microscopic and biochemical examination. Spirulina powder is fine uniform powder, dark green in color, with mild seaweed taste and with no decayed or bitter taste or smell.

Minimum Nutritional Content

Component	Amount	Analysis Method
1. Protein	55 %	AOAC
2. Total Carotenoids	300 mg/100g	AOAC modified
3. Chlorophyll-a	900 mg/100g	AOAC modified
4. Phycocyanin	8,000 mg/100g	DIC method
5. Vitamin B-12	200 mcg/100g	AOAC (microbio. assay)
6. gamma-linolenic acid	900 mg /100g	AOAC

Moisture. Acceptance criteria for each production lot:

1. Moisture	less than 7%	AOAC

Bacteriological Assays. Acceptance criteria for each lot:

1. Standard Plate Count	less than 200,000/g	FDA Bacteriological Manual
2. Molds	less than 100/g	FDA Bacteriological Manual
3. Yeast	less than 40/g	FDA Bacteriological Manual
4. Coliforms	less than 3/g	FDA Bacteriological Manual
5. Salmonella	negative	FDA Bacteriological Manual
6. Staphylococcus	negative	FDA Bacteriological Manual

A.5. Quality and Safety Standards for Spirulina
for the USA Natural Foods Industry

The Natural Products Quality Assurance Alliance (NPQAA)
and The Natural Nutritional Foods Association (NNFA)

Extraneous Materials.
For USA human consumption only, testing of each production lot is
required. *US FDA Guideline acceptance criteria.

1. Insect fragments	*less than150/50g	AOAC (1990) 15th ed.
2. Rodent hairs	*1.0/100g	AOAC 990.09

Heavy Metals. Shown by a typical analysis of spirulina:

1. Lead	less than 2.5 ppm	AOAC
2. Arsenic	less than 1.0 ppm	AOAC
3. Cadmium	less than 0.5 ppm	AOAC
4. Mercury	less than 0.05 ppm	AOAC

Supplementary Guidelines. Shown by a typical analysis of spirulina:

1. No pesticides	4.		No preservatives
2. No herbicides	5.		No stabilizers
3. No dyes	6.		No irradiation

Spirulina finished products

Finished products for human consumption shall meet all relevant USA food
quality and safety standards, and shall follow the appropriate Good
Manufacturing Practice Guidelines.

Minimum Nutritional Content. To be determined.

Moisture. Acceptance criteria for each production lot:

6. Moisture	less than 7%	AOAC

Bacteriological Assays. Acceptance Criteria:

1. Standard Plate Count	less than 200,000/g	FDA Bacteriological Manual
2. Molds	less than 100/g	FDA Bacteriological Manual
3. Yeast	less than 40/g	FDA Bacteriological Manual
4. Coliforms	less than 3/g	FDA Bacteriological Manual
5. Salmonella	negative	FDA Bacteriological Manual
6. Staphylococcus	negative	FDA Bacteriological Manual

Product shelf life:
Producers of finished products shall determine nutrient statements on
labels based on both bulk spirulina powder analysis and nutrient changes
due to tableting and bottling and package shelf life.

A.6. World Spirulina Production 1975-1999

Estimates are based on a world survey of researchers and producers 1993-1996.

Year	Burma	Chile	China	Cuba	India	Japan	Mexico	Taiwan	Thailand	Vietnam	USA-CA	USA-HI	Total
1975	0	0	0	0	0	0	20	0	0	0	0	0	20
1976	0	0	0	0	0	5	45	0	0	0	0	0	50
1977	0	0	0	0	0	11	65	4	0	0	0	0	80
1978	0	0	0	0	0	20	145	4	1	0	0	0	170
1979	0	0	0	0	0	20	200	9	50	0	1	0	280
1980	0	0	0	0	0	20	245	14	50	0	1	0	330
1981	0	0	0	0	0	30	250	19	50	0	1	0	350
1982	0	0	0	0	0	35	250	25	60	0	20	0	390
1983	0	0	0	0	0	45	250	25	60	0	50	0	430
1984	0	0	1	0	0	47	250	60	75	0	55	2	490
1985	0	0	1	0	0	53	250	60	100	1	55	10	530
1986	0	0	1	0	1	60	250	60	110	3	55	20	560
1987	0	0	3	0	3	60	250	60	110	4	70	40	600
1988	0	0	3	0	3	60	250	80	110	4	70	50	630
1989	0	0	8	0	6	50	250	80	110	6	70	60	640
1990	0	0	8	0	7	35	250	90	120	0	120	80	710
1991	4	1	8	0	7	20	250	90	120	0	160	100	760
1992	12	4	12	0	12	20	250	90	120	0	160	120	800
1993	15	5	20	2	20	20	225	90	120	3	160	120	800
1994	20	5	50	10	80	20	100	80	130	5	210	160	870
1995	25	7	120	20	150	20	0	50	150	8	370	250	1170
1996	30	20	250	40	250	20	0	60	150	10	480	400	1710
1997	30	50	500	50	350	20	150	70	150	20	500	450	2340
1998	30	70	700	60	500	20	300	80	150	20	700	500	3130
1999	30	90	900	70	600	20	400	80	150	20	800	600	3760

Year	Burma	Chile	China	Cuba	India	Japan	Mexico	Taiwan	Thailand	Vietnam	USA-CA	USA-HI	Total
1975-99	196	252	2585	252	1989	731	4895	1280	2246	104	4108	2962	21600
% total	1%	1%	12%	1%	9%	3%	23%	6%	10%	0%	19%	14%	100%

Appendix B
The Origins of Earthrise
The Early History of Spirulina Cultivation in the USA

The foundations were begun 20 years ago. Larry Switzer, a visionary bioneer and catalyst, founded the progenitor, Proteus Corporation, in 1976 to develop spirulina blue-green algae as a world food resource. Proteus was funded by a group of private California investors, committed to the vision spirulina represented. Joined by Robert Henrikson, current President of Earthrise, this team began cultivation in the late 1970s.

Hope for the World and its Children

Larry Switzer had been looking for new solutions. He discovered microalgae was 20 times more productive as a protein source than any other food. It could be grown with unused land and water. It was possible to cultivate a pure culture on a large scale in many places around the world.

B.1. Larry and the 'little spiral'.

Scientists discovered spirulina was a safe food, had been consumed for hundreds of years by traditional peoples, and showed promising nutritional, even therapeutic, health benefits. If this blue-green algae were cultivated and consumed by millions of people, it would have tremendous benefits especially for the world's children and our planet's future. It seemed to be a solution we needed.

However, it was all theory – it hadn't been done yet! No one had yet successfully cultivated spirulina on a large scale, produced it as a safe food, and convinced anyone they should indeed eat algae! If it was an idea whose time had come, it was, nonetheless, a daunting task.

Just Close Your Eyes And Chew
By Eric Perlman

B.2. Spirulina article by Eric Perlman, San Francisco Examiner, 1977.

167

Larry first tried to convince the government of Nigeria to build spirulina farms along Lake Chad where it was already growing naturally. But at that time, Nigerians were preoccupied with developing oil. They questioned if anyone in America was eating this algae. If not, why should poor Africans eat it? They were right.

Clearly, to promote algae to the world's hungry people, we first needed to demonstrate its remarkable health benefits to the developed world, and establish it as a valuable new food for everyone.

Our First Prototype Farm in the California Desert

With this in mind, Larry and Robert left the comforts of Berkeley for California's hot Imperial Valley in early 1977. This desert climate was ideal for growing spirulina, and was far from urban pollution. We were joined by a lake ecologist from Berkeley, Dr. Alan Jassby, who helped design the early systems and cultivation programs.

B.3. Our trailer.
B.4. The farm site, with bottles of live spirulina culture.

B.5. Growing up the culture in small trays and wading pools.
B.6. Celebrating the inoculation of a larger pond.

However, innovative projects are not easy, and this was no exception. For three years this small entrepreneurial team sweated in the desert to build a successful farm model. Just about everything that could go wrong, did go wrong, and every problem had to be overcome.

In August 1977, our first farm was totally wiped out by Hurricane Doreen! This unusual August storm caused a one in a hundred year flood, and our little farm was in the middle of ten square miles submerged under 4 feet of water!

B.7. The Flood of '77: On a raft and diving under water to retrieve pumps.

Should we cultivate spirulina in algae farms – or harvest blue-green algae from lakes?

Was this flood a message? Naturally, this setback caused us to reconsider our strategy. If growing spirulina was going to be so difficult, why not just harvest wild blue-green algae from lakes, where it's already growing? We investigated the feasibility of lake harvest schemes. Once again, we redoubled our efforts to cultivate spirulina, for three reasons:

1. Spirulina could be scientifically controlled for a safe product. Natural lakes had mixed blooms of algae throughout the year which we could not control. Some blue-green algae, like some plants, are toxic. We were concerned about producing a contaminated food, and became even more convinced that we needed a scientifically controlled spirulina blue-green algae farm to assure safe food.

2. Spirulina production could improve world food and environmental problems. Harvesting wild algae from lakes would not likely develop into a global business capable of changing world food problems. Cultivating spirulina sustainably and ecologically meant producing food at phenomenally high growth rates. We believed if early bioneers were successful, algae farming would spread all over the world. This would have tremendous economic and environmental impact.

3. Spirulina could gain the support of the world scientific community. Scientists have minimal interest in potentially toxic blue-green algae, such as microcystis, anabaena or aphanizomenon flos-aquae. In contrast, over the past 30 years there has been an explosion of published scientific studies on safe spirulina and chlorella. We realized without the support of world scientists to show the medical community reputable research, algae would not become an accepted world food. In fact, almost all reported medical and health claims made for blue-green algae are based on 20 years of scientific research on spirulina.

Building the Second Prototype Farm

We asked our investors for more funding, relocated, and started a second farm. Ron Henson showed up to help with construction, and eventuallally became the Operations Manager of the current Earthrise Farms. Today he is a Vice-President of Earthrise and Sales Manager of the Animal Health Global Sales.

B.8. Second farm in 1979.
B.9. Larry Switzer, Bruce Carlson, Ron Henson, Robert Henrikson.

B.10. Ron working on a paddlewheel station.
B.11. Attaching paddlewheel blades.

B.12. A paddlewheel circulates water around a pond.
B.13. Tasting wet spirulina freshly harvested.

B.14. Harvested paste has a mild taste.
B.15. Fresh paste and spray dried powder.

We developed larger growing ponds, tested harvesting and drying operations, and felt confident that we could build a commercial size farm. Then we began looking for a new funding source, several million dollars. To interest investors, we now had to prove we could actually sell algae to someone. We began importing spirulina, and developed a partnership with a Japanese company that had just begun growing it in Thailand.

Fortunately we were not alone.
Many other algae bioneers emerged around the world.

A new planetary idea often incarnates through many messengers. This was true with spirulina. While Earthrise was underway in California, other companies began cultivation around the world. Hubert Durand-Chastel, now a Senator of France, encouraged a Mexican company to set up a farm in Lake Texcoco in the 1970s.

Israeli, Indian and European scientists began cultivation research. Others developed village scale and appropriate technology farms, notably Dr. Ripley D. Fox of France, and Dr. C.V. Sesahdri of India. Other bioneers emerged in their respective countries.

In Thailand in 1980, A Japanese company, Dainippon Ink & Chemicals (DIC), built one of the first farms. This global company with $10 billion in sales and a commitment to developing microalgae for food, biochemicals and pharmaceuticals, was led by visionaries who were fascinated with its potential. The former President, Shigekuni Kawamura and current President, Takemitsu Takahashi, are long time spirulina sponsors, and funded development. Heading up the program was "Mr. Spirulina" in Japan, Hidenori Shimamatsu of DIC Bio Division.

Introducing Algae to the Natural Food Market

In 1978, we chose the name Earthrise as our trademark and symbol. Apollo astronauts witnessed the first Rising of the Earth from the moon's surface in 1969. This powerful image represents our awakening to the miracle of our living planet. We rediscovered that blue-green algae, the original photosynthetic life form, offers remarkable health benefits to ourselves, our society and our planet today. We dedicated these gifts to a new awareness of our Earth Rising.

B.16. First Earthrise Logo.

B.17. Earthrise® tablets were first introduced 1979.
B.18. Earthrise people, 1981.

We established Earthrise as a sales company directed by Robert Bellows and Terry Cohen of Boulder, Colorado. Earthrise® began appearing in natural food stores in 1979. Spirulina gained popularity fast. In 1981, the National Enquirer pronounced it a magic diet pill, and consumer demand exploded overnight. Many diet pill companies jumped on the bandwagon, and sold spirulina diet pills that didn't even contain any. There wasn't much real spirulina yet being grown.

The diet boom faded by 1983, but the market began to grow again by 1987. More people continued to experience health benefits from spirulina. More published scientific research documented its therapeutic benefits. As President since 1981, I sold the Earthrise sales company to DIC in 1988 to begin a dynamic new growth phase. With a injection of financial resources, we rapidly expanded. Earthrise® trademark products are now sold in 40 countries, making it the world's best selling spirulina.

The First U.S. Production: Earthrise Farms

We built a relationship with people in DIC, as we imported from their farm in Thailand for Earthrise products. This unusual partnership between California entrepreneurs and Japanese corporate intrapreneurs blossomed. We shared a common vision of spirulina's coming impact on the world economy. Together we founded Earthrise Farms, the first US production farm, as a joint venture in 1982.

B.19. Earthrise Team, 1984: Bellows, Henrikson, Ota, Hamada, Carlson, Shimamatsu, Jassby.

In following years, as funding was needed for expansion, Earthrise Farms became a subsidiary of DIC. Under the leadership of Yoshimichi

Ota, President since 1984, the farm expanded in size, and quality assurance. Dr. Amha Belay from Ethiopia, land of spirulina lakes, is the Senior V.P., and has been responsible for cultivation, and research. Juan Chavez, V.P. and Production Manager, expanded the production capacity to 500 tons a year by 1996.

B.20. Chavez, Ota, Belay, 1996.

B.21-24. Earthrise Farms expansion showing new pond area: 1985, 88, 90, 95.

It's been quite a journey over 20 years, from buckets of algae to the world's largest spirulina farm as a dream became reality.

Bibliography and references

1. Rediscovery of a 3.5 billion year old lifeform

1. Lovelock, James. *The Ages of Gaia.* W.W. Norton&Co., NY, 1988, p. 74.
2. Lovelock, p. 81.
3. Lovelock, p. 115.
4. Lovelock, p. 96.
5. Lovelock, p. 116.
6. Swimme, Brian. *The Universe is a Green Dragon.* Bear & Company, Santa Fe, NM, 1984, p. 137-8.
7. Kavaler, Lucy. *Green Magic: Algae Rediscovered.* Thomas Crowell, NY, 1983, p. 99-101.
8. Kavaler, p. 109-110.
9. Kavaler, p. 95-97.
10. Jassby, Alan. *Spirulina: a model for microalgae as human food.* Algae and Human Affairs. Cambridge University Press, 1988, p. 152.
11. Jassby, p. 153.
12. Ciferri, Orio. *Spirulina, the Edible Organism.* Microbio. Reviews, Dec. 1983, p. 572.
13. Ciferri, p. 578.
14. Switzer, Larry. *Spirulina, The Whole Food Revolution.* Bantam, NY, 1982, p. 12.

2. A nutrient rich super food for super health

1. FDA Talk Paper, No. 41,160, June 23, 1981, US Food and Drug Admin.
2. Jassby, Alan. *Nutritional and Therapeutic Properties of Spirulina.* Proteus Corp, 1983.
3. Switzer, Larry. *Spirulina, The Whole Food Revolution.* Bantam, NY, 1982, p. 22.
4. Jassby, Alan. *Nutritional and Therapeutic Properties of Spirulina.* Proteus Corp., 1983.
5. *The Complete Book of Vitamins and Minerals for Health.* ed. by Prevention Magazine. Rodale Press, Emmaus, PA, 1988, p. 149.
6. Jassby, Alan. *Spirulina: a model for microalgae as human food.* Algae and Human Affairs. Cambridge University Press,1988, p. 158.
7. Jassby, Alan. *Nutritional and Therapeutic Properties of Spirulina.* Proteus Corp. 1983.
8. Kataoka, N. and Misaki, A. *Glycolipids isolated from spirulina maxima.* Agric. Biol. Chem. 47 (10), 2349-2355, 1983.
9. Venkataraman, L.V. and Becker, E.W. *Biotechnology & Utilization of Algae- The Indian Experience.* Sharada Press, Mangalore, India, 1985, p 114-115.
10. Challem, Jack Joseph. *Spirulina. A Good Health Guide.* Keats Publishing, New Canaan CT, 1981, p. 15.
11. Challem, p. 13.
12. Shimamatsu, H. Personal communication, May 8, 1989.

3. Self-care programs with clean green energy

1. Switzer, Larry. *Spirulina, The Whole Food Revolution.* Bantam, NY, 1982.
2. Beasley, Sonia. *The Spirulina Cookbook: Recipes for Rejuvenating the Body.* University of the Trees Press, Boulder Creek, CA, 1981.
3. Howard, Saundra, Dr. *The Spirulina Diet.* Lyle Stuart, Secacus NJ, 1982.
4. Byrne, Kevin, MD. *Cut Your Cholesterol- Now!* Self Care, Nov.-Dec. 1988, p. 27-31.
5. Mckenna, Jeffrey and Shea, John. *Americans are beginning to get NCI's cancer prevention message.* FDA Consumer, Apr.1988, p. 22.

6. National Research Council. *Diet, Nutrition and Cancer.* National Academy Press, Washington DC, 1982.
7. Switzer, Larry. Spirulina, *The Whole Food Revolution.* Bantam Books, NY, 1982, p. 39.
8. Hills, Christopher, Phd. *Rejuvenating the Body Through Fasting With Spirulina Plankton.* Univ. of the Trees, Boulder Creek, CA, 1979, p. 9.
9. Hills, p. 22.
10. Switzer, Larry. *Spirulina, The Whole Food Revolution.* Bantam, NY, 1982, p. 40.
11. Gray, Robert. *The Colon Health Handbook. New Health Through Colon Regeneration.* 11th Ed. Emerald Publishing, PO Box 11830, Reno NV, 1986, p. 13.
12. Gray, p. 36-7.
13. Gray, p. 40.
14. *Spirulina: For a high intensity workout.* Flex, Sep. 1984, p. 61.
15. Bellows, Robert. *Spirulina: Nature's Magic Algae.* Muscle and Fitness, 1983, p 83-4.
16. Garcia, A.M. *Spirulina makes me ten times stronger, like spinach for Popeye.* Gramme International. Sep. 96. p. 11.
17. *Spirulina & Pregnancy.* Body & Soul. Vol. VIII, No.13, p. 8.
18. *Spirulina for Nursing Mothers.* Ballarpur Industries, Ltd., Thapur House, 124, Janpath, New Delhi, India.
19. Dainippon Ink & Chemicals, Inc. *Customer Survey,* 1988.
20. *Nutrition of the Future: Foods that Fight Aging.* Business Week, Oct. 9, 1988, 125-6.

4. New research reveals health benefits

Latest scientific research: effects on AIDS virus, cancer and the immune system

1. Ayehunie, S., Belay, A. et al. *Inhibition of HIV-1 replication by an aqueous extract of spirulina (arthospira platensis).* 7th IAAA Conference, Knysna, South Africa, April 17,1996.
2. Hayashi, T. et al. *Calcium Spirulan, an inhibitor of enveloped virus replication, from a blue-green alga spirulina.* Am. Chemical Soc. and Am. Soc. of Pharmacognosy. Journal of Natural Products, 1996. Vol 59, No. 1, p. 83-87.
3. Hayashi, K. et al. *An extract from spirulina is a selective inhibitor of herpes simplex virus Type 1 penetration into HeLa cells.* Phytotherapy Research, Vol. 7, p.76-80, 1993.
4. Hayashi, O. et al. *Enhancement of antibody production in mice by dietary spirulina.* J. of Nutritional Sciences and Vitaminology, 40, p. 431-441, 1994.
5. Belay, A., Ota, Y. et al. *Current knowledge on potential health benefits of spirulina.* Journal of Applied Phycology. 1993. 5:235-241.
6. Carmichael, W. et al. *The toxins of cyanobacteria.* Scientific American, Jan. 1995.

Cholesterol reduction

7. Nayaka, N. et al. *Cholesterol lowering effect of spirulina.* Tokai Univ, Japan. Nutrition Reports Int'l, June 1988, Vol 37, No. 6, 1329-1337. Nakaya, N. *Effect of spirulina on reduction of serum cholesterol.* Tokai Univ. Progress in Med. Nov. 1986, Vol 6, No. 11.
8. Becker, E.W. et al. *Clinical and biochemical evaluations of spirulina with regard to its application in the treatment of obesity.* Inst. Chem. Pfanz. Nutrition Reports International, April 1986, Vol. 33, No. 4, p. 565.
9. Devi, M.A. and Venkataraman, L.V. *Hypocholesterolemic effect of blue-green algae spirulina platensis in albino rats.* Nutrition Reports International, 1983, 28:519-530.
10. Kato, T. and Takemoto, K. *Effects of spirulina on hypercholesterolemia and fatty liver in rats.* Saitama Med. College, Japan. Japan Nutr Foods Assoc. Jour. 1984, 37:321.
11. Iwata, K. et al. *Effects of spirulina on plasma lipoprotein lipase activity in rats.* Journal Nutr. Sci. Vitaminol. 1990, 36:165-171.

Natural beta carotene and cancer prevention

12. Peto, R. et al. *Can dietary beta carotene materially reduce human cancer rates?* Nature, 1981, 290:201-208.
13. Shekelle, R.B. et al. *Dietary Vitamin A and risk of cancer in the Western Electric study.* Lancet, 1981, 8257: 1185-1189.
14. Menkes, et al. *Serum beta carotene, vitamins A and E, selenium, and the risk of lung cancer.* Johns Hopkins. N.E. Journal of Medicine, Nov. 1986, p. 1250.
15. Blot, William. *Journal of the National Cancer Institute.* Sept. 15, 1993.
16. National Research Council. *Diet, Nutrition and Cancer.* National Academy Press, Washington DC, 1982.
17. Ben Amotz, A. Presentation to *Polysaccharides from microalgae workshop,* Duke University, 1987.

Anti-cancer tumor effects

18. Schwartz, J., Scklar, G., Suda, D. *Inhibition of experimental oral carcinogenesis by topical beta carotene.* Harvard School of Dental Med. Carcinogenesis, May 1986, 7(5) 711-715.
19. Schwartz, J., Scklar, G., Suda, D. *Growth, inhibition and destruction of oral cancer cells by extracts from spirulina.* Cancer & Nutrition, 6/88.
20. Babu, M. et al. *Evaluation of chemoprevention of oral cancer with spirulina.* Nutrition and Cancer V. 24, No. 2, p.197-202, 1995.

Phycocyanin enhances the immune system

21. Iijima, N., Shimamatsu, H., et al. (inventors; Dainippon Ink & Chemicals assignee). *Anti-tumor agent and method of treatment therewith.* US patent pending, ref. P1150-726-A82679, App. 15 Sep. 1982.
22. Dainippon Ink & Chemicals and Tokyo Kenkyukai (inventors and assignee). *Anti-tumoral agents containing phycobilin- also used to treat ulcers and hemorrhoidal bleeding.* 1983, JP 58065216 A 830418.
23. Zhang Cheng-Wu, et. al. *Effects of polysaccharide and phycocyanin from spirulina on peripheral blood and hematopoietic system of bone marrow in mice.* Second Asia-Pacific Conf. Ibid, April, 1994.

Polysaccharides enhance the immune system

24. Besednova, T. et. al. *Immunostimulating activity of lipopolysaccharides in blue-green algae.* Zhurnal Mikrobiologii, Epidemiologii, Immunobiologii, 56 (12), p. 75-79. 1979.
25. Baojiang, G. et. al. *Study on effect and mechanism of Polysaccharides of spirulina on body immune function improvement.* Second Asia-Pacific Conference on Algal Biotechnology. Singapore, April, 1994, p. 24.
26. Zhang Cheng-Wu, et. al. *Effects of polysaccharide and phycocyanin from spirulina on peripheral blood and hematopoietic system of bone marrow in mice.* Second Asia Pacific Conf. Ibid, April, 1994.
27. Qishen, P. et. al. *Endonuclease activity and repair DNA synthesis by polysaccharide of spirulina.* Acta Genetica Sinica (Chinese J. of Genetics). V.15(5) p. 374-381. 1988.
28. Lisheng, L. et. al. *Inhibitive effect and mechanism of polysaccharide of spirulina on transplanted tumor cells in mice.* Marine Sciences, Qindao, China. N.5, 1991, p.33-38.
29. Qureshi, M. et al. *Spirulina extract enhances chicken macrophage functions after in vitro exposure.* J. Nutritional Immunology, V.3(4) 1995, p. 35-45.
30. Qureshi, M. et al. *Dietary spirulina enhances humoral and cell-mediated immune functions in chickens.* Immunopharmacology and Immunotoxicology, 1996 (submitted).
31. Qureshi, M. et al. *spirulina exposure enhances macrophage phagocytic function in cats.* Immunopharmacology and Immunotoxicology, 1996 (submitted).

32. Belay, A., Henson, R., Ota, Y. *Potential Pharmaceutical Substances from Aquaculturally Produced Spirulina.* Earthrise Farms, Calipatria CA. Presented to World Aquaculture Society, New Orleans. Jan. 1994.

Sulfolipids stop the AIDS virus

33. Gustafson K, et al. *AIDS-Antiviral sufolipids from cyanobacteria (blue-green algae).* Journal of the National Cancer Institute, August 16, 1989, p 1254.
34. Kataoka, N. and Misaki, A. *Glycolipids isolated from spirulina maxima: structure and fatty acid composition.* Agric. Biol. Chem. 47 (10), 2349-2355, 1983.
35. Venkataraman, L.V. and Becker, E.W. *Biotechnology & Utilization of Algae- The Indian Experience.* Sharada Press, Mangalore, India, 1985, p 114-115.
36. Boyd, M, Gustafson, K. et al. *Cyanovirin-N, a novel HIV-inactivating protein that targets viral GP120.* Am. Soc. of Pharmacognosy, UC Santa Cruz. July 27-31, 1996.

Reduces kidney poisons from mercury and drugs

37. Yamane, Y. *The effect of spirulina on nephrotoxicity in rats.* Presented at Annual Sym. of the Pharmaceutical Society of Japan, Apr. 15, 1988, Chiba Univ, Japan.
38. Fukino, H. et al. *Effect of spirulina on the renal toxicity induced by inorganic mercury and cisplatin.* Eisei Kagaku. 1990, 36:5.

Effects against diabetes and hypertension

39. Takai, Y. et al. *Effect of water soluble and insoluble fractions of spirulina over serum lipids and glucose resistance of rats.* J. Jap. Soc. Nutr. Food Sci. 1991, 44: 273-277.
40. Iwata, K. et al. *Effects of spirulina on plasma lipoprotein lipase activity in rats.* Journal Nutr. Sci. Vitaminol. 1990, 36:165-171.

Builds healthy lactobacillus

41. Tokai, Y., et al. *Effects of spirulina on caecum content in rats.* Chiba Hygiene College Bulletin (Japan), Feb. 1987, Vol. 5, no. 2.
42. Archer, D.L. and Glinsmann, W.H. *Intestinal infection and malnutrition initiate AIDS.* US FDA. Nutrition Research, 1985. 5:19-19. Archer, D.L. and Glinsmann, W.H. *Enteric infections and other cofactors in AIDS.* Immunology Today, 1985, Vol. 6, no.10.

Wound healing and antibiotic effects

43. Clement, G. et al. (inventors; Institute Francais de Petrol, assignee.) *Wound treating medicaments containing algae.* Fr. M. 5279 (Int. Cl. A61k), 11 Sep. 1967.
44. Yoshida, R. (inventor). *Spirulina hydrolyzates for cosmetic packs.* Japan. Kokai 77 31, 838 (Int. Cl. A61k7100), 10 Mar. 1977.
45. Martinez-Nadal, N.G. *Antimicrobial activity of spirulina.* Paper presented at X Intl Cong. of Microbiology, Mexico City, Aug. 1970.
46. Jorjani, G., Amirani, P. *Antibacterial activities of spirulina platensis.* Maj. Iimy Puz. Danisk.Jundi Shap, 1978, 1:14-18.

Benefits for malnourished children

47. Ramos Galvan, R. *Clinical experimentation with spirulina.* Colloque sur la valeur nutritionelle des algues spirulines, Rueill, May 1973. Nat. Inst. of Nutrition, Mexico City.
48. Sautier, C. and Tremolieres, J. *Food value of spirulina in humans.* Ann. Nutrition Alim, 1976, 30:517-534. (French).
49. Fox, Ripley D. *Integrated village health and energy system, Farende, Togo.* April 4, 1986 letter. Fox, R. D. Algoculture: Spirulina, hope for a hungry world. Pub. by Edisud, Aix-en Provence, France, 1986.

50. Seshadri, C.V. *Large Scale Nutritional Supplementation with spirulina alga.* All India Project. Shri Amm Murugappa Chettiar Research Center (MCRC) Madras. 1993.
51. Annapurna, V. et al. *Bioavailability of spirulina carotenes in preschool children.* National Institute of Nutrition. Hyderabad, India. J. Clin. Biochem Nutrition. 10 p. 145-151. 1991.
52. Fica, V. et al. *Observations on the utilization of spirulina as an adjuvant nutritive factor in treating some diseases accompanied by a nutritional deficiency.* Clinica II Medicala, Spitalui Clinic Municipiului, Bucuresti. Med Interna 36 (3), 1984.
53. Miao Jian Ren. *Spirulina in Jiangxi China.* Academy of Agricultural Science, Jiangxi province, China. Paper presented at Soc. Appl. Algology, Lille, France, Sep. 1987.
54. Yonghuang, W. et al. *The study on curative effect of zinc containing spirulina for zinc deficient children.* Shenzhen Blue-Green Algal Biotech. Corp, Guangdong. Capital Medical College, Beijing, 5th Int'l Phycological Congress, Qingdao, China. June 1994.

Iron bioavailability and correction of anemia

55. Johnson P., Shubert, E. *Availability of iron to rats from spirulina, a blue-green alga.* Nutrition Research, 1986, Vol. 6, 85-94.
56. Takemoto, K. *Iron transfer from spirulina to blood in rats.* Saitama Medical College, Japan, 1982.
57. Takeuchi, T. *Clinical experiences of administration of spirulina to patients with hypochronic anemia.* Tokyo Medical and Dental Univ., Japan, 1978.

GLA and prostaglandin stimulation

58. Jassby, Alan. *Nutritional and Therapeutic Properties of Spirulina.* Proteus Corp. 1983.
59. Tudge, C. *Why we could all need the evening primrose.* New Scientist, Nov. 1981, 506:23.
60. Kunkel, S.L. et al. *Suppression of chronic inflammation by evening primrose oil.* Progress in Lipids, 1982, Vol. 20, p. 885-888.
61. Kernoff, P.B.A, et al. *Antithrombotic potential of DGLA in man.* British Med. Journal, 1977, 2:1441-1444.
62. Vadaddi, K.S., Horrobin, D.F. *Weight loss produced by evening primrose oil administration.* IRSC Med. Sci., 1979, 7:52.
63. Huang, Y.S. et al. *Most biological effects of zinc deficiency corrected by GLA.* Atheroscelosis, 1982, 41:193-208.
64. Horrobin, D.F. *The possible roles of prostaglandin E1 and of essential fatty acids in mania, depression and alcoholism.* Progress in Lipids, 1981. Vol 20, 539-541. Horrobin, D.F. *Loss of delta-6-desaturase activity as a key factor in aging.* Med Hypotheses, 1981, 7:1211-1220.
65. Passwater, R.A. *Evening Primrose Oil.* Keats Publishing Co. New Canaan, CT, 1981.
66. Lopez-Romero, D. *Gamma linolenic acid as a base of treatment for infirmities with evening primrose oil and spirulina.* Med. Holistica, Madrid, Spain, 12 Oct. 1987.
67. Hudson and Karlis. *The lipids of the alga spirulina.* J. Sci Food Agric., 1974, 25: 759.
68. Nichols, B., Wood, B. *The occurrence and biosynthesis of gamma linolenic acid in spirulina platensis.* Lipids, 1986, Vol 3, No. 1, 46-50.
69. Roughhan, P. Grattan. *Spirulina: A source of dietary gamma-linolenic acid?* J. Sci. Food Agric., 1989, 47, 85-93.

Weight loss research

70. Becker, E.W. et al. *Clinical and biochemical evaluations of the alga spirulina with regard to its application in the treatment of obesity.* Inst. Chem. Pfanzenphysiologie. Nutrition Reports International, April 1986, Vol. 33, No 4, 565.

Reduces effects of radiation

71. Loseva, L.P. and Dardynskaya, I.V. *Spirulina- natural sorbent of radionucleides.* Research Institute of Radiation Medicine, Minsk, Belarus. Presented at the 6th Intl Congress of Applied Algology, Czech Republic, Sep. 9, 1993.

72. Sokolovskiy, V. *Correspondence from the First Secretary BSSR Mission to the United Nations,* May 20, 1991.

73. Belookaya, T. *Correspondence from the Chairman of Byelorussian Committee "Children of Chernobyl"* May 31, 1991.

74. Qishen, P. et. al. *Radioprotective effect of extract from spirulina platensis in mouse bone marrow cells studied by using the micronucleus test.* Toxicology letters. 1989. 48:165-169.

75. Evets, P. et. al. *Means to normalize the levels of immunoglobulin E, using the food supplement spirulina.* Grodenski State Medical Univ. Russian Federation Commitee of Patents and Trade. Patent (19)RU (11)2005486. Jan. 15, 1994.

5. The variety of products around the world

1. *A safe diet pill, You'll never go hungry.* National Enquirer, June 1, 1981.

2. Henson, R. *Spirulina: Health food for the Aquarium.* Freshwater and Marine Aquarium. 1993. p 70-72.

3. Henson, R. *Spirulina algae improves Japanese fish feeds.* Aquaculture Magazine. Nov/Dec 1990. V. 6 N.6. p 38-43.

4. Henson, R. *Spirulina and Wheatgrass: Super-Foods for birds.* American Cage-Bird Magazine. April, 1993. p. 48-51.

5. Qureshi, M. et al. *Immune enhancement potential of spirulina platensis in chickens.* Dept. of Poultry Science, NC State Univ. Raleigh NC. Poultry Sci. Assoc, Aug. 1994.

6. *Linablue-A (Natural blue colorant of spirulina origin) Technical Information.* Dainippon Ink & Chemicals, Tokyo, Japan, 1985.

7. Kronik, M. and Grossman, P. *Immunoassay techniques with fluorescent phycobiliprotein conjugates.* Clin Chem., 1983, 29:9,1582-86.

8. *A better way to track diseases.* Chemical Week, August 31, 1985.

9. Kawamura, M. et al. 1986. *A new restriction enzyme from spirulina platensis.* Nucleic Acids Research, 1986, Vol. 14, No. 5.

10. Shimamatsu, Hidenori. Personal communication, 1988.

11. Belay, A., Ota, Y. et al. *Current knowledge on potential health benefits of spirulina.* Journal of Applied Phycology. 1993. 5:235-241.

6. How spirulina is ecologically grown

1. Belay, A., Ota, Y. et al. *Production of high quality spirulina at Earthrise Farms.* Second Asia-Pacific Conference on Algal Biotech. Phang et al. eds. Univ. of Malaysia, 1994.

2. Tredici, M. et. al. *Novel photobioreactors for the mass cultivation of spirulina.* Spirulina, Algae of Life. Bulletin de l'Institute Ocean. Monaco. N. special 12. April 1993.

3. Tomaselli, L et. al. *Physiology of stress response in Spirulina.* Spirulina, Algae of Life.

7. Resource advantages and world food politics

1. Lappé, Frances Moore. *Diet For a Small Planet.* Ballantine, N.Y, 1982, p. 85.

2. Lappé, pg. 76.

3. *Rediscovering Planet Earth.* U.S. News and World Report. Oct 31, 1988, pg. 68.

4. Rainforest Action Network. 300 Broadway #28, San Fran, CA 94133.
5. Lappé, Frances Moore. *Diet For a Small Planet.* Ballantine Books, N.Y, 1982.
6. Robbins, John. *Diet for a New America.* Stillpoint Publishing, Walpole, NH, 1988.
7. Repetto, Robert, et. al. *Wasting Assets. Natural Resources in the National Income Accounts.* World Resources Institute. 1989.
8. *California's thirstiest crops.* Water Education Foundation. S.F. Chronicle, March 4, 1991.
9. Grantham, Richard. *Seeking a biological solution for the greenhouse dilemma.* Institut d'Evolution Moleculaire, Univ. Claude Bernard Lyon. Villeurbanne cedex, France, 1988.
10. Ota, Yochimichi. Earthrise Farms. Personal communication, 1988.
11. *Antidote for a smokestack.* Time Magazine. Oct. 24, 1988, p. 72.
12. Wolf, Edward C. *Beyond the Green Revolution: New Approaches for Third World Agriculture.* Worldwatch 73, Washington DC,1986, p. 9.
13. Gardner, Gary. *Asia is losing ground.* Worldwatch. Nov/Dec. 1996. Washington DC,1986, p. 20-21.
14. *An assessment of the resource base that supports the global economy.* World Resources Institute, Basic Books, New York, 1988, p. 4.
15. Lappé, F.M and Collins, J. *Food First: Beyond the Myth of Scarcity.* Houghton Mifflin, Boston, 1977.
16. The Hunger Project, 2015 Steiner St. San Francisco, CA, 94115.
17. Higgins, et. al. *Potential population supporting capacities of lands in the developing world.* Technical Report of Project FPA/INT/513. FAO, United Nations, Rome, 1983.
18. Postel, Sandra. *Land's End.* Worldwatch , May-June 1989, p. 13.
19. Gardner, Gary. *Asia is losing ground.* Worldwatch. Nov/Dec. 1996. Washington DC,1986, p. 19.

8. Spirulina in the developing world

1. U.N. World Health Organization, Geneva, Switzerland. Correspondance. June 8, 1993.
2. Switzer, Larry. *Spirulina- The Whole Food Revolution.* Bantam, NY, 1982, p.115-116.
3. Association Pour Combattre la Malnutrition par Algoculture (ACMA). A non-profit organization. 34190 St. Bauzille-de-Putois, France.
4. Fox, Ripley D. *Algoculture: Spirulina, hope for a hungry world.* Edisud, Aix-en Provence, France, 1986. (in French)
5. Fox, Ripley D. 1984. *Spirulina- the alga that can end malnutrition.* Futurist, Feb. 1985.
6. Seshadri, C.V. *Large Scale Nutritional Supplementation with spirulina alga.* All India Project. Shri Amm Murugappa Chettiar Research Center (MCRC) Madras. 1993.
7. Seshadri, C.V. and Jeeji Bai, N. *Spirulina Nat. Sym..* MCRC, Madras, India 1992.
8. Becker, E.W. *Development of Spirulina Research in India.* Spirulina, Algae of Life. Bulletin de l'Institute Oceanographique. Monaco. N. Special 12. April 1993. p 141-155.
9. Min Thein. *Production of Spirulina in Myanmar.* Spirulina, Algae of Life. Bulletin de l'Institute Oceanographique. Monaco. N. Special 12. April 1993. p 175-178.
10. Fox, Denise. *Health Benefits of Spirulina.* Algae of Life. Bulletin de l'Institute Ocean. Monaco. N. Special 12. April 1993. p 179-186
11. Nguyen Lan Dinh. 1990. In Fox, Denise. *Health Benefits of Spirulina.* p 183.
12. Bucaille, P. 1990. In Fox, Denise . *Health Benefits of Spirulina.* p 183.
13. Picard, M.E. 1993. In Fox, Denise. *Health Benefits of Spirulina.* p 182.
14. Brouillat, E. 1993. In Fox, Denise. *Health Benefits of Spirulina.* p 182.

9. Microalgae's role in restoring our planet

1. Kavaler, Lucy. *Green Magic: Algae Rediscovered.* Thomas Crowell, NY, 1983, p. 60.
2. Oguchi, Mitsuo, et al. *Food production and gas exchange system using blue-green algae (spirulina).* National Aerospace Laboratory, Chofu, Japan. COSPAR, July, 1986.
3. Macelroy, R.D. and Smernoff, D.T. *Controlled ecological life support systems.* Proceedings of 26th COSPAR. Toulouse France, July, 1986. NASA TM 88215, 1987.
4. Robinson, K. S. *A Colony in the sky. Mission to Mars.* Time, Sept. 23, 1996. p.49.
5. Todd, Nancy Jack and Todd, John. *Bioshelters, Ocean Arks and City Farming.* Sierra Club Books, San Francisco, 1984, p. 118.
6. Todd, p. 155.
7. Fox, Ripley. *Spirulina, production & potential.* Edisud. Aix-en-Provence, France. 1996.
8. Jourdan, Jean-Paul. *Cultivez Votre Spiruline - Manuel de Culture Artisanale de la Spiruline.* 1996 (in print). To order: Jean-Paul Jordan, Le Castanet, Mialot, 30140 Anduze, France. Fax: 33 4 66 85 02 39.
9. Jassby, Alan. *Spirulina: a model for microalgae as human food.* Algae and Human Affairs, Cambridge Univ. Press, 1988, p. 171.
10. Fox, Ripley. *Spirulina, production & potential.* Edisud. Aix-en-Provence, France. 1996.
11. *Blue-green algae for rice production.* FAO of the U.N. FAO Soils Bulletin. Rome, 1981.
12. Metting, Blaine. *Micro-algae in agriculture.* Micro-Algal Biotechnology, ed. by Bororwitzka. Cambridge Univ., Cambridge, U.K., 1988.
13. Grantham, Richard. *Seeking a biological solution for the greenhouse dilemma.* Institut d'Evolution Moleculaire, Univ. Claude Bernard Lyon, Villeurbanne cedex, France.1988.
14. Perlman, David. *Scientists turn ocean into green new world.* San Francisco. Chronicle, Oct. 10, 1996, p. A2
15. Lovelock, James. *The Ages of Gaia.* W.W. Norton & Co. Inc., N.Y. 1988, p. 179.
16. Carol Cone, Cone Comm. Cone/Roper Survey. Business for Social Responsibility *Corporate Social Responsibility.* 1030 15th St. NW, Washington, DC. April, 1994.

Appendix A: Quality and safety standards

1. Delpeuch, F. et. al. *Consumption as food and nutritional composition of blue-green algae among populations in the Kanem region of Chad.* Ann. Nutr. Aliment. 29, 497-516. 1976.
2. Institut Francais du Petrol. *Rapport ou Comite Consultatif des Proteines* OAA/ OMS/ FISE- Etat d'Avancement du Procede IFP de Production d'algues Dec 1970, p. 11.
3. Fadoul, L. *Les algues bleues du Kanem.* Rapport de mission par L. Fadoul, A. Avrem et G. Le Guedes (experts de la division de la nutrition, FAO) Juin 1971.
4. Takemoto, K. *Subacute toxicity study with rats.* Saitama Medical College, Japan, 1982.
5. Atatsuka, K. *Acute toxicity and general pharmacological studies.* Meiji College of Pharmacy, Japan,1979.
6. Chamorro-Cevallos, G. *Toxicological research on the alga spirulina.* UNIDO, 24 Oct. 1980, UF/MEX/78/048. (French)
7. Becker, W. E., Vanattaraman, L.V. et al. *Production and utilization of the blue-green algae Spirulina in India.* Biomass, 4, 105-125, 1984.
8. Chamorro, G. et al. *Subchronic toxicity study in rats fed Spirulina.* J. de Pharmacie de Belgique 43, 29-36, 1988.
9. Salazar, M., Chamorro, G. et al. *Effect of spirulina consumption on reproduction and peri- and postnal development in rats.* Food and Chemical Toxicity 34, 353-359. 1996.
10. Waslein, C. et al. *Uric acid levels in men fed algae and yeast as protein sources.* J. Food. Sci. 1970, 35, 294-8.

Bibliography and References

11. Jassby, Alan. *Spirulina: a model for microalgae as human food.* Algae and Human Affairs, Cambridge Univ Press, 1988, p. 159.
12. Boudene, C. et al. *Evaluation of long term toxicity on rats with spirulina.* Ann Nutr. Aliment., 1976, 30: 577-588.
13. Til, H.P. and Williams, M. *Sub-chronic toxicity study with dried algae in rats.* Cent. Inst. for Nutrition and Food Research, Zeist, Ned., 1971.
14. Fevrier, C. and Seve, B. *Incorporation of spirulina into pig diets.* Ann. Nur. Aliment., 1976, 29: 625-30 (French).
15. Earthrise Farms. *Five year testing of heavy metals in spirulina, 1983-1987.* 1988.
16. Carmichael, W.W. *The toxins of Cyanobacteria.* Sci. American, Jan. 1994, p. 78-86.
17. An, J., and Carmichael, W.W. *Technical Booklet for the Microalgae Biomass Industry: Detection of microcystins and nodularins using an enzyme linked immunosorbant assay (ELISA) and a protein phosphate inhibition assay (PPIA).* Dept. Bio. Sci. Wayne State Univ, Dayton OH. July 1996.

Resource guide

Spirulina (selected references)

1. Belay, A., Ota, Y., Miyakawa, K., and Shimmatsu, H. *Production of high quality spirulina at Earthrise Farms.* Phang et al. Algal Biotech in Asia-Pacific Region, p. 92-102. 1994.
2. Fox, Ripley D. *Spirulina, production and potential.*1996. To order: Editions Edisud, La Calade, R.N.7, 13090 Aix-en-Provence, France. Fax: 33 42 21 56 20.
3. Fox, Ripley D. *Algoculture: Spirulina, hope for a hungry world.* Edisud, Aix-en Provence, France, 1986.
4. Laboratoire de la Roquette. c/o Dr. Ripley D. Fox, 34190 St. Bauzille-de-Putois, France.
5. Richmond, Amos. *Spirulina.* Micro-Algal Biotechnology. ed. by Borowitska. Cambridge Univ. Press, Cambridge, UK, 1988.
6. Vonshak, A. and Richmond A. *Mass Production of the blue-green algae spirulina: an overview.* Biomass, 1988, p. 233-247.
7. Jassby, Alan. *Some public health aspects of microalgal products.* Algae and Human Affairs, ed. by Lembi and Waaland. Cambridge Univ. Press, 1988, p. 181-202.
8. Venkataraman, L.V. and Becker, E.W. *Biotechnology & Utilization of Algae- The Indian Experience.* Sharada Press, Mangalore, India.

Other References

Hawken, Paul. *The Ecology of Commerce.* Harper Collins, New York. 1993.
Ausubel, Kenny. *Seeds of Change: The Living Treasure.* Harper San Francisco. 1994.
Wilson, Edward O. *The Diversity of Life.* W.W. Norton, New York. 1992.
Utne Reader Magazine. Minneapolis, MN.
Worldwatch Institute. State Of The World. 199. W.W. Norton & Company. NY, 1993.
WorldWatch Magazine. Worldwatch Institute, Washington, DC 20036.
World Resources Yearbook. World Resources Inst., Basic Books, NY, 1993.
Food First. Institute for Food and Development Policy, 145 Ninth St., San Francisco, CA.
National Resources Defense Council (NRDC), 40 West 20th St., New York, NY 10011.
Greenpeace, 1436 U Street N.W., PO Box 3720, Washington, DC 20007.
Conservation International, 1015 18th Street, N.W., Suite 1000, Washington, DC, 20036.
Rainforest Action Network, 300 Broadway #28, San Francisco, CA 94133.
The Hunger Project. Ending Hunger: An Idea Whose Time Has Come. Praeger, NY, 1985.
Gaia, An Atlas of Planet Management. ed by Dr. Norman Myers. Anchor Books, Doubleday, Garden City, NY, 1984.
Berry, Thomas. *The Dream of the Earth.* Sierra Club, San Francisco, 1988.

Index

Index

International Editions

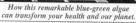

How this remarkable blue-green algae
can transform your health and our planet.

Robert Henrikson

Robert Henrikson

M I C R O A L G A

SPIRULINA

Superalimento del futuro

Por qué la microalga
spirulina puede
transformar su salud
y nuestro planeta

URANO

HRANA ZA ZEMLJANE

Spirulina

*ali o tem, kako ta nenavadna
modrozelena alga lahko spremeni
vaše zdravje in naš planet-Zemljo*

Robert Henrikson

地球食品
施柏健螺旋藻

本书由世界著名的
螺旋藻生产商美国 EARTHRISE 总裁撰写
是螺旋藻领域最权威的著作

［美］ 罗伯特·夏礼逊 著

186

ORDER FORM
Send to: Ronore Enterprises, Inc.
PO Box 1017, Petaluma, CA 94953-1017 USA

Yes, please send this book:

Earth Food *Spirulina*
**How this remarkable blue-green algae
can transform your health and our planet.**

Quantity		Discount %	per Book	Amount
_____	1 book	0 %	$ 13.95	_____
_____	2 - 4 books	20 %	$ 11.20	_____
_____	5 - 9 books	30 %	$ 9.80	_____
_____	10-24 books	40 %	$ 8.40	_____
			Total for Books	_____

California residents please add 7.5% sales tax _____

USA: shipping: $2.00 first book
$1.00 each additional book _____

International: shipping: $4.00 first book
$ 2.00 each additional book _____
I can't wait 3-4 weeks for surface Book Rate

USA, Canada, Mexico: $ 4.00 first book for air mail
$ 3.00 each additional book _____

International: $ 7.00 first book for air mail
$ 5.00 each additional book _____

AMOUNT ENCLOSED (US funds) _____

Name _____

Address _____

City, State, Zip _____

Thank you for your order.